Teaching Biology

to **KS4**

Mark **Winterbottom**

non-specialist **handbook**

Series Editor: Elaine **Wilson**

Hodder & Stoughton

A MEMBER OF THE HODDER HEADLINE GROUP

We are grateful to the following examination boards for permission to reproduce questions: Edexcel, OCR, NEAB, NICCEA, SEG and WJEC

Order queries: please contact Bookpoint Ltd, 130 Milton Park, Abingdon,Oxon OX14 4SB. Telephone: (44) 01235 827720, Fax: (44) 01235 400454. Lines are open from 9.00–6.00, Monday to Saturday, with a 24-hour message answering service.

A catalogue record for this title is available from The British Library

ISBN 0 340 75356 0

First published 1999
Impression number 10   9   8   7   6   5   4   3   2
Year   2005   2004   2003   2002

Copyright © 1999 Mark Winterbottom

Cover photo from Oxford Scientific Films
Typeset by Wearset, Boldon , Tyne and Wear.
Printed in Great Britain for Hodder & Stoughton Educational, a division of Hodder Headline Ltd, 338 Euston Road, London NW1 3BH by The Bath Press, Bath.

Contents

iii

4

Variation, Inheritance, Evolution and Classification $\qquad$ 156

5

Living Things in their Environment $\qquad$ 185

Examination questions annotated with
specimen answers $\qquad$ 211

1 Life Processes and Cell Activity

Life and its characteristics

BACKGROUND

Biology is the study of life. The question 'What is life?' is difficult to answer; however, we can distinguish seven characteristics which are shared by all living organisms: movement, sensitivity, nutrition, respiration, excretion, reproduction and growth.

Pupils from KS2 know that plants and animals share the following characteristics: nutrition, growth and reproduction. They also know that animals can move. They may not appreciate what life is, or that any of these processes are characteristic of life. Pupils at KS3 and KS4 double and single award should realise that a living organism shows all the seven characteristics outlined above.

Pupils' understanding of 'life' is very varied. Teaching will be successful if you allow pupils to express their misconceptions and then challenge those misconceptions with specific examples. Research has highlighted some problems in learning about life:

- Younger pupils often attribute human emotions to inanimate objects, which suggests that they consider such objects to be alive.
- Many younger pupils will label something living or non-living only according to the most obvious characteristics, usually movement, nutrition or growth. For example, they could label a candle, car or an inflating balloon as being alive.

Research suggests that teaching should focus on the way in which living things are unified by the possession of the seven characteristics of life. Bear in mind that pupils often leave lessons with an understanding of life 'for' biology lessons, while their everyday understanding remains unaltered.

KEY STAGE 3 CONCEPTS

The characteristics of life

Pupils will find it helpful to understand the derivation of the word 'biology'. They should be able to link *bio-* with 'life' and, given the range of *-ology* subjects, are likely to realise this means the 'study of'.

To assess pupils' preconceptions, set up a circus of (i) animal, plant and microbe pictures/specimens, (ii) some 'non-alive' man-made objects e.g. pencils, pens, furniture, and (iii) a lighted candle, a toy car, a robot, a teddy bear and running water (which are often mistakenly classified as living). Then develop pupils' understanding in three stages.

1 Compare animals with man-made, non-living objects (pupils find it much easier to identify characteristics of life in animals than in plants), bringing out those characteristics possessed by animals which make them living. Ensure you include both terrestrial and aquatic, and vertebrate and invertebrate examples of animals. Ask pupils to identify the characteristics of life in themselves. You could use this exercise to include a brief summary of each characteristic, including whereabouts in the body each occurs in humans.

2 Compare animals with plants, identifying how plants demonstrate each of life's characteristics. Pupils find this more difficult; helpful hints are given below.

- *Movement* – remind pupils that plants move in response to light; show them a plant which has been on the window-sill for a few days. You can demonstrate movement in a venus fly-trap (*Dionaea muscipula*) – just stroke the hairs inside the trap several times.
- *Respiration* – many pupils will find respiration hard to appreciate if they have not yet studied it. Tell them it is the release of energy from food, and discuss why plants may need energy, e.g. flowering, fruiting and growth.
- *Sensitivity* – plants must be sensitive to light; if they were not sensitive, they could not move towards it.
- *Growth* – have some pre-germinated cress seeds on some cotton wool. Pupils can measure the length of the shoots in subsequent lessons.
- *Reproduction* – show pupils a spider plant which has grown runners to

make a new plant (asexual reproduction). Remind them that plants also make seeds which grow into new plants (sexual reproduction).

- *Excretion* – many pupils will know that plants give out oxygen as a waste material which humans inhale (and use for respiration).
- *Nutrition* – pupils may find the idea of photosynthesis hard to appreciate if they have not studied it. Tell them that plants make their own food using light energy from the Sun (by photosynthesis). Many less able pupils may know that plants take in some minerals through their roots.

  Pupils often have difficulty identifying seeds and trees as being alive, indeed some will not classify them even as plants! This may be because of their tough outer layer which differs so much from a normal green appearance. Planting the seeds and watching them grow often helps pupils realise that they are plants and that they are living.

3 Compare living things with non-living examples which are often classified by pupils as living, e.g. flame, running water, cars, bicycles, teddy bears and robots. Focus on helping pupils to realise that to be living, something must show all seven of life's characteristics, not just one or two.

Having completed these activities, you should ensure that each pupil has concluded correctly which processes are characteristic of living things. To help them remember, use the acronym: MRS GREN, or ask them to make up their own rhymes.

Movement
Reproduction
Sensitivity
Growth
Respiration
Excretion
Nutrition

Assessing pupils' learning

- Pupils could design an artificial living organism from man-made materials. They must include all seven of life's characteristics and write an account explaining why they have included specific features in that organism.
- Set up a new circus of different machines, fire and familiar animals and plants (or pictures of them). Pupils should record the presence or absence

of the seven characteristics of life and state which of the specimens are alive.

- Ask pupils to identify the characteristics of life in some household machines. They should not find all seven in any machine.
- Pupils could write about how they would investigate whether life exists on Mars.

Key Stage 4 concepts

Pupils at KS4 would benefit from recapping the work from KS3. Depending on the ability of the group, you could condense or expand upon the activities suggested above. Provide some more information about each of life's processes in animals and plants – you can find such information in the relevant chapters. Do not try to teach the whole National Curriculum in one lesson. Stress to pupils that most of sections 2 and 3 of the Sc2 National Curriculum is concerned with studying life's processes.

chapter 2

Cells, tissues, organs and organ systems

BACKGROUND

All living things are composed of cells: the basic unit of life. Bacteria (prokaryotes) and protista are unicellular – made from just one cell. Humans and green plants are multicellular – comprising more than one cell. Animal and plant cells share the following structures: a cell membrane (which controls what enters and leaves the cell), the cytoplasm (a solution of food and waste materials, and a site in which chemical reactions can take place) and a nucleus (which contains the genetic material – the information which runs the cell). In addition, plant cells are surrounded by a cell wall (a strong, but flexible case which gives the cell strength), and include chloroplasts (which contain chlorophyll, a green pigment which traps light energy during photosynthesis) and a large vacuole (a food store which helps support the cell). Animal cells do contain vacuoles but they are very small.

In multicellular organisms, different groups of cells are often specialised for different functions. A collection of cells which have the same function is a tissue (e.g. a muscle). Tissues may be grouped together into organs (e.g. the heart), and organs into organ systems (e.g. the circulatory system). There are six major organ systems in the human body, each of which maintains one or more of life's characteristics: digestive, circulatory, gaseous exchange, excretory, endocrine and nervous.

Pupils will be unfamiliar with the cell at KS2. At this stage there is emphasis on large scale processes whereas KS3 and 4 relate such processes to the cells, tissues, organs and organ systems which carry them out.

In both KS3 and 4, pupils must know the components of plant and animal cells and the functions of those components. At KS4 double and single award, pupils must recognise the similarities and differences between plant and animal cells. Because cells cannot be seen with the naked eye, you will also need to introduce pupils to the microscope.

At KS3, pupils must understand the definitions of tissue and organ, and recognise which organs are responsible for each of life's characteristics in plants and animals. At KS4 double and single award, pupils must also be familiar with each organ system. Pupils' understanding of organ systems is developed throughout the programme of study.

KEY STAGE 3 CONCEPTS

The microscope

Although not included explicitly at KS3 or 4, pupils must be able to use a microscope. Begin by asking pupils to make their own magnifiers.

- Make a hole in a milk bottle top about 2 mm wide and place a small drop of water in the hole.
- If pupils put their eye very close to the drop and the specimen just under the hole, the object should be magnified by up to 100 times.

Before pupils touch the microscopes themselves, stress that they are fragile and delicate. They should always be carried with two hands: one underneath and one holding the curved part of the body. They should not touch the lenses. More able pupils should be familiar with the names of the parts of the microscope (you can find a diagram in most textbooks). Some such names are easy to remember:

- The specimen is viewed on the stage; just as actors are viewed on the stage.
- The objective lens is near the object being viewed.
- The eyepiece lens is near the eye.

To view a specimen, it must be prepared on a microscope slide. The specimen must be thin enough for light to pass through, otherwise no light will reach your eye and the specimen will appear black.

- Place the specimen on the slide and add a drop of water or stain using a pipette or a tap.
- Slowly lower a cover slip onto the drop using a mounted needle (Figure 2.1). This method stops air bubbles forming when the liquid spreads out. Tell pupils that the cover slip prevents liquid touching the objective lens which, although it can be cleaned, should be kept dry. Ask your technician to clean the lens if necessary; using paper towels will scratch the lens.

- Cover slips should be handled by the edge to prevent fingerprints which may obscure the specimen.
- If there is too much liquid under the cover slip, some may leak out. Place filter paper against the side of the cover slip; excess water will be sucked up by capillary action.

S **Safety Advice:** Cover slips are thin and easily breakable. If using a stain, check you know its safety hazards.

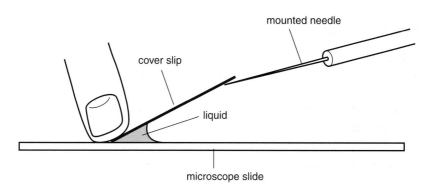

Figure 2.1 *Lowering a cover slip onto a specimen*

Having prepared a slide, you can examine the specimen:

- Rotate the lens of lowest magnification (usually marked ×10) into position (until it is vertical).
- If your microscope has a built-in lamp, switch it on. If not, face a window (but not direct sunlight) or bench lamp and use the mirror to shine light through the stage and up to the eyepiece (look down the eyepiece and move the mirror until the field of view appears white).
- Place the slide on the stage under the clips, so the specimen is directly beneath the objective lens.
- Look at the microscope from the side, at the level of the stage, and use the focussing knob to move the objective lens as close as possible to the slide. During focussing, the objective lens or the stage may move up and down; microscope designs differ.
- Focus upwards (so the distance between objective lens and stage increases) until the specimen comes into focus. Never focus downwards when looking through the eyepiece; you may crash into the slide.
- To examine the specimen under high power, focus under low power and then rotate the next objective lens into position. When rotating the lens, check it does not crash into the slide.

Pupils should practise drawing specimens. Drawings should be large (at least half a page), should not include the outline of the field of view (i.e. the drawing should not be bounded by a circle) and should have a heading. More able pupils at KS3 and 4 can include the magnification (the ratio of the apparent size to the actual size) – multiply the objective lens magnification by the eyepiece magnification.

KEY STAGE 4 CONCEPTS

The microscope

Pupils should be familiar with microscopes and preparing slides. Remind pupils of good practice. Persistent mistakes include (i) holding the microscope with one hand, (ii) focussing down onto the slide and breaking it, (iii) turning the objective lens round and crashing into the slide, and (iv) forgetting to mop up excess liquid.

KEY STAGE 3 CONCEPTS

The structure of animal and plant cells

Pupils sometimes find it difficult to appreciate how cells, atoms and molecules are related to each other. Because cells and atoms both have nuclei, pupils equate them in their minds. Accordingly, although pupils may identify objects as being made of atoms and molecules, they may suggest that cells are an alternative to atoms or molecules in living things. Pupils may even suggest that biological molecules are made of cells!

More able pupils could draw plant and animal cells, either from prepared microscope slides, or from slides they have prepared themselves. Less able pupils could label pre-prepared diagrams.

- To prepare onion cells for observation, peel off the epidermis from one layer of an onion. The epidermis is the thin translucent layer which you sometimes see when cutting an onion. Lay the epidermis on a slide and add iodine solution. Cover with a cover slip.
- Alternatively, use Canadian pond weed (*Elodea canadensis*) leaves which are thin enough to observe cells; these may have more chloroplasts than onion cells. Stain with iodine and view as above.
- To prepare slides of cheek cells, pupils should use a new cotton bud, rub it along the inside surface of their mouth and dab it onto a slide. Add a drop of 1% methylene blue and cover with a cover slip.

S **Safety Advice:** Check DfEE advice about preparing cheek cells. If pupils prepare them, place used cotton buds in absolute ethanol. If you cannot prepare cheek cells and have no prepared slides of animal cells, use liver cells (see Burton 1999). Alternatively, stick clear tape to the wrist, peel it off suddenly and stick it to a microscope slide; some cells may be visible.

Research suggests that pupils do not appreciate the three-dimensional nature of cells, especially when presented with 2D images from textbooks and the microscope. Try making a model of an animal cell by filling a plastic bag with water (the cytoplasm), including a small ping-pong ball (the nucleus) and sealing it. Don't worry if the bag changes shape – this happens in a normal animal cell. Place the model on an overhead projector. It will project a 2D image onto the screen, helping pupils to see how the 2D and 3D images relate. To model a plant cell, take the animal cell, and add some small chunks of green polystyrene to represent the chloroplasts. Place the bag in a sandwich box to represent the cell wall.

Depending on time, you can either tell pupils the functions of the different parts of the cell, or ask them to research them from textbooks or CD-ROMs.

Assessing pupils' learning

- Carry out a verbal test on the parts of the cell and their functions.
- Pupils could produce leaflets about each part of the cell, or about plant or animal cells as a whole.
- Pupils could make their own models of plant and animal cells, indicating the functions of each part.
- More able pupils could write newspaper articles, explaining the structure and function of plant and animal cells.

KEY STAGE 4 CONCEPTS

The structure of animal and plant cells

Reintroduce pupils to the structure and function of cells. They must recognise those structures which are shared by plant and animal cells. Take note that:

- Pupils sometimes mistakenly label the cell sap in the vacuole as the cytoplasm.
- Pupils sometimes label the vacuolar membrane or the cell wall as the cell membrane. Because the cell membrane in plant cells is often pushed up against the wall, it is often invisible under the microscope. To make it visible, prepare onion cells in concentrated salt solution. The membrane will pull away from the cell wall.

KEY STAGE 3 CONCEPTS

Tissues, organs and organ systems

Your department should have pre-prepared slides or textbooks with diagrams of different cell-types, making up different tissues. Ask pupils to relate the structure of these cells to their function:

- Palisade cells in a leaf contain many chloroplasts to do lots of photosynthesis.
- Root hair cells protrude from the root, increasing the surface area across which water and mineral salts can be absorbed.
- Sperm cells have tails which allow them to swim to the egg in the human oviduct.
- Egg cells (ova) are large and contain plenty of food to support the embryo.
- Ciliated epithelial cells (cells with tiny hairs called **cilia**) are used to remove mucus from the trachea (windpipe); the hairs sweep mucus to the throat.
- Red blood cells are *biconcave* to give them more surface area to absorb oxygen. They also lack a nucleus, giving more space to store haemoglobin which carries the oxygen.
- White blood cells are specialised to fight disease; they can engulf invading microbes and produce antibodies.
- Nerve cells are long and thin to carry messages around the body.

Explain to pupils that a group of different tissues forms an organ (e.g. muscle and nervous tissue comprise the heart), each of which has a particular function and helps the body to carry out particular processes of life. If you have a model human body in your department, you could gradually take it to pieces, naming each organ, and asking which characteristic of life each one is involved in. Pupils could research the functions of different organs.

- Less able pupils can match up pre-prepared definitions of organs, their function, and the life process(es) with which each is involved.

KEY STAGE 4 CONCEPTS

Tissues, organs and organ systems

In addition to the above, KS4 pupils should know that a group of organs together (such as the bladder and kidneys) comprises an organ system (e.g. the excretory system). Pupils should be able to identify organ systems (digestive, circulatory, gaseous exchange, nervous, endocrine and excretory), the organs which make up these systems, and decide which of life's characteristics each organ system is responsible for.

- Use a model human body to introduce the organ systems.
- Ask pupils to research a particular organ system and present their findings to the class.
- Ask pupils to label the particular organs which comprise each organ system on prepared diagrams.

Ensure you refer to the breathing system or the gaseous exchange system. Do not call it the respiratory system. Many pupils arrive from KS2 thinking that respiration is gaseous exchange. It is not. Respiration is the process by which cells release usable energy from food by oxidation. The gaseous exchange system provides the cells with oxygen for this process (see Chapter 5). More able pupils could discuss why we have specialised organ systems. One reason is that they provide greater efficiency in carrying out life's processes.

11

Assessing pupils' learning

- Pupils could describe how the organs in each organ system are involved in each characteristic of life.
- They could play a game of 'body patience' (Shaw 1997).
- Less able pupils could research one particular organ system, collating results into a class poster describing all the organ systems in the body.

References

Burton, I.J. (1999) A simple technique for preparing liver cells for microscopical examination and a description of their structural features. *Journal of Biological Education* **33**:113–114

Shaw, A. (1997) Body patience. *School Science Review* **79**:106–107

chapter **3**

Chromosomes and cell division

BACKGROUND

The nucleus of every cell contains DNA (deoxyribonucleic acid) – the encoded instructions (genetic information) which run the cell. DNA is a double helix of nucleotides (pentose sugar, phosphate group and organic base). The length of DNA which encodes a particular instruction is called a gene. A string of genes, packed together with protein, makes up a chromosome. There are two versions of every chromosome in adult human cells. Such cells are called diploid. These versions of chromosomes are similar but not identical. For example, in a particular place on each of a pair of chromosomes, there may be a gene which determines eye colour. On one chromosome, the version of the gene may denote blue eye colour; on the other, it may denote brown eye colour. Such different versions of genes are called alleles.

All cells divide. There are two types of cell division: mitosis and meiosis. Mitosis produces two cells (diploid cells) with exact copies of their parents' chromosomes. Mitosis is used by unicellular organisms to reproduce, and by multicellular organisms to reproduce, grow and repair damaged tissue. Meiosis produces two cells (haploid cells) with half the diploid number of chromosomes: the sex cells or gametes. These only have one version of every chromosome. Female gametes are called egg cells or ova (in animals and plants). Male gametes are sperm cells (in animals) and pollen cells (in plants). When two sex cells unite in fertilisation, this restores diploidy to the resulting embryo.

Mitosis is the method by which organisms maintain their large number of cells. Meiosis (which halves the chromosome number in the cell) is necessary because two cells (the egg and sperm) are going to fuse to form a zygote (a normal cell). Because adult cells must have two sets of chromosomes (be diploid) to survive, the gametes must be haploid when they fuse together.

Research has found that teaching mitosis and meiosis side by side leads to confusion. The processes are so similar that pupils find it difficult to distinguish between them. If you have a choice, introduce mitosis in the context of growth. Deal with meiosis later as an introduction to variation and inheritance (Chapters 19 and 20), and then compare it to mitosis to highlight the similarities and differences.

KEY STAGE 4 CONCEPTS

Chromosomes and genes

This is the first stage at which chromosomes and genes are introduced. Tell pupils that different species' cells have different numbers of chromosomes. Humans have 46 in each cell (except for sperm and eggs which have 23). Less able pupils often mistakenly imagine that a human has 46 chromosomes in their whole body; correct this misconception before it takes root.

You should introduce two different 'models' of chromosome structure. The first is functional and depicts chromosomes as strings of genes.

- Your department may have 'poppit' beads or model-making kits to demonstrate this.
- Use a piece of rubber tubing to model the chromosome whilst talking to pupils. Wrap pieces of different coloured tape around the tubing to denote genes, and include lots of genes along the length of the chromosome.

Tell pupils that genes are the instructions for running the cell and controlling features of the body. Because genes are normally discussed in the context of specific features, e.g. a gene for eye or hair colour, some pupils may think that different cells contain different genes. For example, eye cells contain eye colour genes, whereas skin cells contain skin colour genes. This is wrong: all human cells contain all 46 chromosomes. Therefore, all cells contain copies of all genes. In different tissues, genes are simply switched on or off as required.

The second model of a chromosome is structural. This is the model of a chromosome you will find in most textbooks and mirrors the appearance of chromosomes under the microscope.

Assessing pupils' learning

- Pupils could model this using two pipe cleaners, twisted around each other at the centre to make the centromere. This whole structure is now referred to as the chromosome.

- Most biology textbooks have labelled diagrams of chromosomes. Your department may have prepared slides which pupils can examine, draw and label.
- Pupils could make models of chromosomes using pipe cleaners.
- Pupils could label a prepared diagram of a chromosome.

Chromosomes, and therefore genes, are lengths of DNA. More able pupils will need to appreciate how the structure of DNA allows the genes to carry encoded information. Liken the double helix of DNA to a twisted ladder. Bases pair up (adenine with thymine, guanine with cytosine); such a pair comprising a rung of the ladder. The order in which bases are arranged linearly in the DNA strand acts like a code.

Your department may have a DNA modelling kit, or paper stencils of phosphates, sugars and bases to cut out and stick together in the correct order. Jackson (1996) describes a home-made DNA modelling kit.

Pupils can extract DNA from kiwi fruit. The method is described below:

- Add 25 g sodium chloride to 80 ml washing-up liquid and make up the volume with water to 1 litre.
- Stir to dissolve the sodium chloride but try to avoid frothing.
- Add 100 ml of this solution to a cut and mashed kiwi fruit in a 250 ml beaker.
- Incubate in a water bath at 60°C for 15 minutes and filter the resultant green mush through coffee filter paper until you have collected about 4 cm depth of filtrate in a test tube.
- Run ice-cold methylated spirits *slowly* down the side of the test tube so it forms a 1 cm depth layer above the filtrate.
- At the interface between the two layers, white strands of DNA will form.

S **Safety Advice:** Methylated spirits is highly flammable. Extinguish naked flames and keep stock bottles closed when not in use. Warn pupils not to eat the kiwi fruit.

Immediately after cell division, the chromosome consists of one double-helical strand of DNA wrapped around protein molecules (histones). This whole structure is coiled up like a telephone wire. The chromosomes uncoil before the next cell division; while uncoiled, they make an exact copy of themselves which remains joined to the original at the centromere. When the cell starts the process of division, the chromosomes coil up again becoming shorter and fatter. At this stage, the chromosome is made up of two identical chromatids. You can demonstrate this coiling using a piece of rubber tubing or thick rope (Busby 1995). Ask a pupil to hold one end, and then twist the other end. The tubing will coil up in the same way as a chromosome.

Mitosis

Mitosis is not included at KS3 so is new to all pupils at this level. Start by explaining that mitosis is a form of cell division which produces cells which are genetically identical to their parent cell. It is important for two reasons:

- *Growth and repair:* cells divide, each daughter cell increases in size, and then each of these divides again. Ensure you stress that the daughter cells do increase in size. Many teachers simply say that cell division causes growth. However, less able pupils may imagine one cell of a standard size, simply being cut in half over and over again and not getting any bigger (i.e. no growth occurring).
- *Asexual reproduction:* unicellular organisms (such as bacteria or amoebae) can reproduce by splitting themselves in half by mitosis. Some plants reproduce asexually and can grow new plants by mitosis. Show pupils a spider plant as an example.

Pupils often fail to realise why cells produced for growth and repair must be genetically identical. For example, if a muscle cell is dividing, it must produce another muscle cell. If the genetic content of the cell were to change, the cell would not have the instructions to do its job properly. Tell more able pupils that in animals, mitosis occurs all over the body. In plants, it only occurs in meristematic cells in the buds and in the growing tip of the shoot and root.

More able pupils must understand the mechanism of mitosis. You can find diagrams of each stage in most textbooks. Although pupils are unlikely to need to remember the names of the stages, they must be able to label the spindle, chromosomes, chromatids, centromeres and centrioles. Stress the important points:

- The chromosomes replicate (copy themselves) to form two chromatids joined at the centromere.
- The chromosomes become shorter and fatter and line up on the equator of the cell.

- Centrioles (which have moved to each end of the cell) help to construct a spindle. This consists of lots of protein fibres which grow toward the chromosomes and fix onto them.
- The spindle fibres contract (like muscles) and pull the chromatids apart. Each end of the cell gets one copy of the genetic material within each chromosome. The nuclear membrane reforms around these new chromosomes.
- In animal cells, the cell membrane pinches in, so dividing the two new cells. In plants, a new cell wall grows across the middle of the cell to separate the two daughter nuclei.

Assessing pupils' learning

- Give pupils diagrams of each stage of mitosis and ask them to annotate the diagrams using textbooks.
- Pupils can make a 2D model of a dividing cell at a particular stage of mitosis. Pipe cleaners serve very well as chromosomes, and string can be used for the spindle. If different groups do different stages, you could model the whole process.
- Pupils can make a 'flick-book' showing the cell at different stages of mitosis.

Meiosis

This is the first stage at which meiosis is introduced to pupils. Discuss why it is important: it produces gametes for sexual reproduction. When two gametes fuse, they produce the first cell of a new organism. All body cells have two versions of each chromosome; each gamete providing one version of each chromosome. Meiosis divides normal cells into gametes which contain only one chromosome of each pair. If gametes were produced by mitosis, when the sperm and egg fused together in fertilisation, the resultant zygote would have too many chromosomes to survive.

To show what happens during meiosis, provide pupils with diagrams. Like mitosis, the chromosomes replicate during interphase into two identical copies. However, in the first meiotic division, chromosomes line up in pairs on the equator of the cell and the spindle fibres pull a whole chromosome to each end of the cell. The cell itself then divides into two daughter cells, containing half the number of chromosomes. Each of these daughter cells then divides in a similar way to mitosis. The four daughter cells resulting from the two meiotic divisions therefore only have one chromosome of each pair. When a sperm cell fertilises an egg cell, a full set of chromosomes is present in the resulting zygote.

Pupils often get confused between pairs of chromosomes, and the two copies of the genetic material present (in the two chromatids) within each chromosome just before cell division. You will only avoid such problems if you use clear diagrams to explain the process. You could challenge any confusion by asking pupils to compare explicitly the structure of one replicated chromosome with a pair of chromosomes.

Assessing pupils' learning

- Use questions to establish whether pupils understand why meiosis is important.
- Give pupils diagrams of each stage and ask them to annotate them using textbooks.
- Pupils can make a model of a dividing cell at a particular stage of meiosis.
- Pupils can identify the similarities and differences between meiosis and mitosis in a table. You will find such compare and contrast tables in many textbooks.
- Pupils can make a 'flick-book' showing the stages of meiosis.

References

Busby, S. (1995) A practical way of learning about DNA using rubber tubing as a model. *Journal of Biological Education* 29:95–97

Jackson, M.E. (1996) A kit for making models of DNA. *School Science Review* 78:114–115

chapter 4

Diffusion, osmosis and active transport

BACKGROUND

Diffusion is the mechanism by which most substances are transported around living organisms. It is the net movement of a substance from a region of high concentration to a region of lower concentration. Much transport of chemicals in animals and plants involves molecules entering or leaving cells through a semi-permeable membrane (i.e. the cell membrane). Some substances can simply diffuse across the membrane, for example small ions and lipid-soluble molecules. Diffusion of water across a semi-permeable membrane is given a special name – osmosis. That is, the diffusion of water from a region of high water concentration to a region of lower water concentration across a semi-permeable membrane. An alternative definition is the diffusion of water from a region of low solute concentration to a region of high solute concentration across a semi-permeable membrane. The final membrane-transport method is active transport. Organisms use this method to move into cells molecules which cannot diffuse across the membrane unaided, either because they are insoluble in lipid (the main constituent of most membranes), or because they would have to diffuse against a concentration gradient. Active transport requires energy in the form of adenosine triphosphate (ATP).

Transport of molecules around living organisms is not included explicitly at KS3 or at KS4 single award. At KS4 double award, you will probably need to cover diffusion, osmosis and active transport at relevant points in the remainder of the programme of study.

KEY STAGE 4 CONCEPTS

Diffusion

Diffusion is responsible for the majority of transport of materials within living organisms. For example:

- Food molecules diffuse from the gut cavity into blood capillaries.
- Oxygen molecules diffuse from the air sacs (alveoli) in the lungs into capillaries.
- Urea diffuses from the blood of a foetus into the mother's blood in the placenta.
- Carbon dioxide diffuses from leaf air spaces into mesophyll cells.

To demonstrate diffusion, carefully place a crystal of potassium manganate(VII) in a beaker of very still water. The potassium manganate(VII) will dissolve into solution and spread (diffuse) through the beaker. Having confirmed that diffusion is the net movement or spreading of a material from a region of high to lower concentration (down a concentration gradient), you should explain what is happening at a particulate level. To do so, remind pupils about basic kinetic theory:

- substances are made up of particles.
- the particles move around at random when in gaseous form, liquid form or in solution.
- the particles can collide and will bounce off each other.

Use a tray of balls or marbles to represent particles. Place one set of coloured balls at one end of a tray, and different coloured balls at the other end of the tray. If you give the particles energy by vibrating the tray gently and evenly, the two sets of balls will mix. That is, there will be net movement of balls of each colour from a region of their high concentration to lower concentration.

Many pupils will try to attribute almost decision-making powers to particles. They often imply that particles look around them, decide they are in a region of high concentration and move to a region of lower concentration. To explain why this is incorrect, you must explain the term *net* movement. Repeat the tray experiment: number each ball and ask each pupil to focus on just one ball. They will realise that each ball actually rolls at random all around the tray. The overall effect of this is that each different coloured set of balls spreads out evenly.

In many cases, diffusion occurs across membranes. To explain this, consider the membrane to be like a sieve. If there are more molecules on one side of the membrane, there is more chance of a molecule randomly bouncing through a hole to the other side. When there is an equal number of particles on each side of the membrane, there is an equal chance of molecules bouncing in both directions.

Osmosis

Many syllabuses only include osmosis for higher level pupils. All abilities of pupil find osmosis difficult and their understanding will depend upon how you introduce the topic. Begin by stating that osmosis is the net movement of water molecules from a region of high water concentration to a region of lower water concentration. This may confuse some pupils because they are unlikely to have considered water concentration before. Simply remind them that water is made up of particles and so can diffuse in the same way as a solute.

Diffusion of water is only called osmosis if the water diffuses across a semi- or partially-permeable membrane. This happens when water enters or leaves cells through the cell membrane. Such a membrane will only allow certain substances through and not others.

Demonstrate a semi-permeable membrane with a simple model (Figure 4.1).

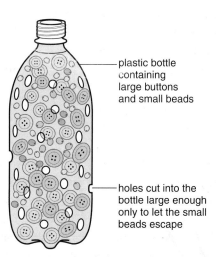

plastic bottle containing large buttons and small beads

holes cut into the bottle large enough only to let the small beads escape

Figure 4.1 *A model of a semi-permeable membrane*

Use large buttons for sugar molecules and small beads for water molecules. When you give the particles energy by shaking the bottle, the water molecules pass out of the bottle, whereas the sugar molecules are kept within it.

The easiest way to record this is by a simple picture (Figure 4.2). The concentration of water, relative to solute, is high on the left-hand side and lower on the right-hand side of the membrane. Therefore, water diffuses from the left-hand side to the right-hand side by osmosis. Although there is also a concentration gradient of solute, the molecules cannot fit through the membrane and so stay on the right-hand side.

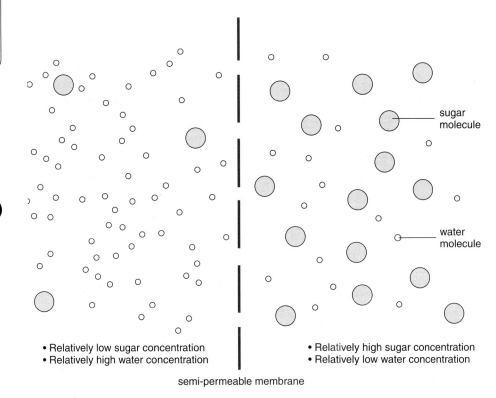

- Relatively low sugar concentration
- Relatively high water concentration

- Relatively high sugar concentration
- Relatively low water concentration

sugar molecule

water molecule

semi-permeable membrane

Figure 4.2 *A schematic diagram of a semi-permeable membrane. The water molecules are small enough to fit through the membrane, whereas the sugar molecules are too large*

You must always talk about the *relative* concentration of water. Pupils often say that water moves from a region where there is lots of water, to a region where there is little water; that is incorrect. Imagine placing a cell in a beaker of strong salt solution. There is clearly more water in the whole beaker than in the cell. However, the relative water concentration in the beaker is less than that in the cell. Remember, water moves from a region of high water concentration (relative to solute) to a region of lower water concentration (relative to solute).

Osmosis may appear rather irrelevant until you place it in some context. There are three neat practicals which you can carry out. For many syllabuses, these are best to include when discussing plant transport. Before you start, remind pupils that the cell cytoplasm is a solution of organic molecules and inorganic salts. Diffusion and osmosis can occur into or out of any solution, the cytoplasm being no exception. Remind pupils that the cell wall in plant cells is fully permeable. The cell membrane of both plant and animal cells is semi-permeable.

Demonstrating osmosis

Visking tubing is semi-permeable and can be used to mimic cell membranes. Soak the tubing in water for an hour before use. Cut a 15 cm length of tubing, seal it at one end, three-quarters fill it with sugar solution and seal it at the other end (you can either tie a knot in the tubing itself or tie it with cotton). Opening the tubing takes some dexterity; a sharp pencil is often the only way to get the two sides of the tubing apart. Take another piece of tubing, three-quarters fill it with water and seal. Ensure the liquid levels are identical in each piece of visking tubing. You have made two models of a plant cell (without the cell wall). Place each in its own beaker of water. Because the relative concentration of water in the sugar solution is lower than in the beaker, water will diffuse in by osmosis and the 'bag' of visking tubing will expand. Nothing will happen to the visking tubing containing only water.

How does osmosis affect onion cells?

This practical is suitable for more able pupils. Red onions are used because it is easy to see the red cell contents contract and expand (if red onions are unavailable, use rhubarb epidermis). Cut a red onion into quarters, remove the outer dry layers and peel away the scale leaves (the thick fleshy layers). You need the thin red epidermis from the top of one of these layers; try using fine forceps. If this doesn't work, try cracking one of the layers – you will see the thin translucent red layer between the two halves. Place this layer on a slide, add a drop of distilled water and cover with a cover slip. Leave for five minutes and examine under the microscope. Repeat the experiment, but using 1.0 M sucrose solution.

When an onion cell is in pure water, water diffuses into the cytoplasm because the relative concentration of water within the cell is lower than in the pure water. This pushes the cell membrane up against the cell wall. In this state, the cell is referred to as being turgid. If the cell wall was not there (e.g. in an animal cell), the cell would eventually burst. When an onion cell is placed in sucrose solution, water diffuses out of the cytoplasm. The cell membrane is pulled away from the cell wall; the cell is referred to as being flaccid or plasmolysed. Ask pupils to annotate diagrams of turgid and plasmolysed cells, explaining the difference between them. You can make a model to reinforce the difference (Figure 4.3).

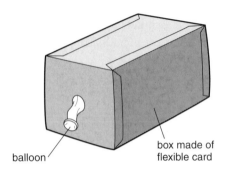

balloon   box made of
flexible card

Figure 4.3 *A model of a plant cell. The cardboard represents the cell wall and the balloon represents the cell membrane. Inflation and deflation of the balloon represents water passing into and out of the cell, respectively*

How does osmosis affect potato cells?

Cut two strips of potato, about 5 mm × 5 mm × 5 cm. Ask pupils to measure them exactly and find their mass in grams (to two decimal places). Place one piece of potato into a test tube and cover with distilled water. Place the other into a test tube and cover with 1.0 M sucrose solution. Leave for 30 minutes. After this time, because the cells in sucrose have become flaccid, the strip of potato is soft and floppy, has shrunk, and has lost mass. By contrast, the potato in the water, whose cells are turgid, is hard and stiff, has expanded and has gained mass.

How does osmosis affect raisins?

Place raisins in a beaker of distilled water for a week. Because the sugar concentration in the cells of the raisins is high, the concentration of water is relatively low. Water diffuses by osmosis into the raisins' cells and the raisins increase in size and bulge out.

> **Assessing pupils' learning**
> - Ask pupils to predict the direction of water flow in a variety of scenarios, including those above.
> - Pupils should write up the practicals and explain their results. Less able pupils should predict the direction of water flow, but need not explain the details of osmosis.

- Pupils should predict the effect of osmosis on animal cells. If there is time, you can investigate this by placing cheek cells in distilled water and in 1 M sucrose solution.

S **Safety Advice:** Check current DfEE advice about using cheek cells. If you cannot prepare cheek cells, a method for preparing animal liver cells is described by Burton (1999).

Active transport

If an organism wants to move a particle from one side of a membrane to another (i) against its concentration gradient, or (ii) when that molecule is too large or not soluble in the fatty component of the membrane, it must invest energy in the form of ATP (adenosine triphosphate), which is produced by respiration within the cells (see Chapter 5). This is called active transport.

A good example of active transport occurs when plants pump mineral salts into their root hair cells. Because the concentration of minerals is higher within the cytoplasm of the root hair cell than in the soil water, they tend to diffuse out of the plant. Ask pupils to imagine a 'waterfall' of minerals flowing over a cliff top. Because minerals are so valuable to the plant, the root hairs do not want to lose them irrevocably. To retrieve the minerals, the plant must pick them up and take them back to the top of the cliff; this needs energy. They use protein pumps in the membranes of the root hair cells to pump the minerals back into the root hair (from a region of low concentration to a region of higher concentration).

Assessing pupils' learning

- Use questions to help pupils summarise active transport. Ensure they record that it requires energy, and that it commonly moves molecules against their concentration gradient.
- Pupils could imagine themselves to be particles moved by active transport and explain what happens as they go across the membrane.
- Pupils could construct a transport board game, including diffusion, osmosis and active transport.

References

Burton, I.J. (1999) A simple technique for preparing liver cells for microscopical examination and a description of their structural features. *Journal of Biological Education* **33**:113–114

2   Humans as Organisms

chapter 5

Respiration

BACKGROUND

Respiration is the process by which cells convert chemical energy in food into a usable form of chemical energy in the cell (adenosine triphosphate, ATP). There are two forms of respiration: aerobic (which needs oxygen to take place) and anaerobic (which does not need oxygen). In aerobic respiration, cells produce ATP by oxidising glucose to form carbon dioxide and water in three distinct sets of reactions, one of which occurs in the cytoplasm, and two of which occur in the mitochondria (small organelles in the cytoplasm). If there is no oxygen available, anaerobic respiration occurs in the cytoplasm. Aerobic respiration produces more ATP than anaerobic respiration. Anaerobic respiration is also known as fermentation. It can take two forms: lactic fermentation in animals and bacteria, and alcoholic fermentation in plants and fungi.

Pupils do not meet respiration at KS2. Aerobic respiration is introduced at KS3 and anaerobic respiration at KS4 double award. Respiration is not included at KS4 single award. Although respiration is introduced at KS3, even pupils at the end of KS4 still fail to realise its importance. They rote-learn that respiration is one of life's characteristics, but do not understand that respiration releases the energy from food.

KEY STAGE 3 CONCEPTS

Does food contain energy?

A good way to demonstrate that food contains energy is to burn it: a crisp or digestive biscuit on the end of a mounted needle work well.

S **Safety Advice:**

- Do not burn nuts because students may have allergies. Check the ingredients of the food you do burn: it may have nuts added for flavouring.
- Warn pupils that some foods will burn with very intense flames; hair should be tied back and ties tucked in.
- Be aware of students with asthma.

Most pupils will tell you that light energy and heat energy are given off, and will realise that the energy came from the food. Stress that the energy in food is chemical energy. Some pupils may think that you put the energy in when you lit the food. This cannot be true because you only held the food in the flame for a few seconds to light it, whereas it continued to burn for a longer period. Many pupils will say that you have 'made' energy by burning the food. Stress that food is an energy store from which energy can be released, just like a battery.

Your next step is to measure how much energy is released. Use the apparatus in Figure 5.1 to do this. Remind pupils to turn off the Bunsen burner having ignited the food, otherwise the water may be heated up by the Bunsen burner as

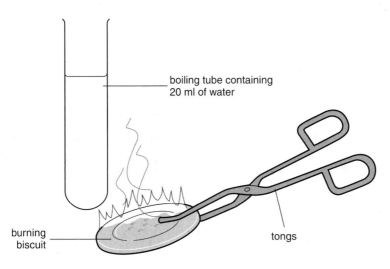

boiling tube containing
20 ml of water

tongs

burning
biscuit

Figure 5.1 *Measuring the energy released by burning a piece of food*

well as by the food. In this experiment, you are utilising the fact that it takes 4.2 J of energy to increase the temperature of 1 ml of water by 1°C.

Even 20 ml of water will be heated up noticeably by burning a small piece of biscuit. Ensure pupils find the mass of the food before burning, and that they position the biscuit at the end of the tweezers. Pupils should measure the temperature of the water before and after heating, and calculate the energy given off per gram of food using the equation:

$$\text{energy (J/g)} = \frac{\begin{array}{c}\text{change in water} \\ \text{temperature (°C)}\end{array} \times \begin{array}{c}\text{volume of} \\ \text{water (ml)}\end{array} \times 4.2 \text{ (J/°C/ml)}}{\text{mass of food (g)}}$$

You could give pupils a range of foods (e.g. crisps, banana chips and digestive biscuits) of which to compare the energy content for an Sc1 investigation. Pupils should predict which food will release the most energy on burning. To do so, they will need to know the proportion of each food that is protein, carbohydrate and fat, and that 1 g of protein releases 17 kJ energy, 1 g of carbohydrate releases 17 kJ of energy, and 1 g of fat releases 38 kJ of energy. More able pupils could calculate the predicted energy released per gram of food. Less able pupils should predict that the food which contains most fat will release most energy. Your department may have a calorimeter which works on the same principle and can be used as a demonstration to measure energy content of food.

Assessing pupils' learning

Pupils should write up the experiments and calculate the energy released by the food. More able pupils should be encouraged to evaluate their experiment in detail.

KEY STAGE 4 CONCEPTS

You can use the above as revision and a reintroduction to respiration. If you are short of time, demonstrate the experiments.

KEY STAGE 3 CONCEPTS

What is needed for food to burn, and what is produced by burning?

Food needs oxygen to burn. Pupils should compare the way food burns in open air, in a limited supply of air (on a combustion spoon in a sealed gas jar), and in a gas jar of pure oxygen. Pupils should observe the intensity of the flame produced and the duration of burning.

S **Safety Advice:**

- Burning food in pure oxygen should be a teacher demonstration.
- Do not burn nuts, or foods containing nuts.
- Be careful of foods burning with intense flames: tie hair back and tuck in ties.

Pupils should conclude that food burns better (i.e. with a more intense flame) when it has more oxygen. It therefore needs oxygen to burn. Relate this to your previous experiment and conclude that food needs oxygen to release energy as light and heat.

You now need to demonstrate that food produces carbon dioxide and water when it burns. Limewater goes cloudy in the presence of carbon dioxide and cobalt chloride paper turns from blue to pink in the presence of water. Pupils should collect two gas jars of the smoke produced by burning a piece of food (Figure 5.2). Add a piece of cobalt chloride paper to one gas jar and 5 cm depth

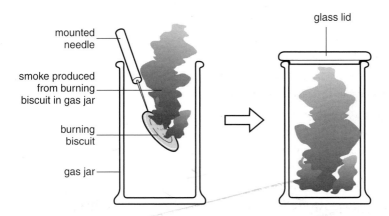

Figure 5.2 *Collecting the products of burning in a gas jar*

limewater to another. The cobalt chloride paper should be handled with tweezers because sweat on pupils' fingers will turn it pink. The limewater, as well as turning cloudy, may turn yellow because of the smoke. Explain that pupils should look for cloudiness, and should ignore any yellow colour formed.

Assessing pupils' learning

Pupils should write up the experiments and conclude that oxygen is needed for food to burn and that carbon dioxide and water are released when food burns. Pupils should conclude that oxygen is needed to release the energy from food. All pupils should write a word equation for burning.

$$\text{Food} + \text{Oxygen} \rightarrow \text{Carbon dioxide} + \text{Water} + \text{Energy}$$

KEY STAGE 4 CONCEPTS

Use the above to reintroduce respiration. Pupils are more likely to remember the word equation for respiration if they see proof that oxygen is needed to release energy from food, and that carbon dioxide and water are formed as by-products.

KEY STAGE 3 CONCEPTS

How is energy released in cells?

Having written the word equation, tell pupils that our cells predominantly use sugar, specifically glucose, as a source of energy. To release energy from sugar, all living cells carry out a process very similar to burning, with the same word equation. That process is respiration:

$$\text{Glucose} + \text{Oxygen} \rightarrow \text{Carbon dioxide} + \text{Water} + \text{Energy}$$

Many pupils believe that breathing is respiration. This is wrong. Breathing is gaseous exchange; release of energy in cells is respiration. Their misconception appears to arise from the different uses of the word respiration in scientific and everyday parlance. For instance: (i) the kiss of life is often referred to as artificial respiration, and (ii) respiratory diseases affect the breathing system. Some pupils will suggest that respiration is *needed* for breathing. This is only correct to the extent that all bodily processes need respiration to release energy to allow them to happen.

Pupils should realise how the reactants are acquired by, and the products expelled from, the body.

- To show that oxygen is taken into the body from inhaled air, (i) collect some inhaled air by putting the lid on a gas jar in the open air, and (ii) collect some exhaled air in a gas jar over water. Measure the time a candle will burn in each gas jar (Figure 5.3). Alternatively, use an oxygen sensor.
- To show that glucose is present in food, carry out a test for sugar (see Chapter 6).
- Most pupils will accept that urine contains water. Water is also expelled in exhaled air. Ask pupils to breathe on blue cobalt chloride paper – it will turn pink in the presence of water. This should be handled with tweezers as moisture from the fingers will turn it pink.

- To show the difference in carbon dioxide content between exhaled and inhaled air, use the apparatus in Figure 5.4.

S **Safety Advice:** Ensure tubes are disinfected before and after use, and that each tube is only used by one person. Be careful of pupils accidentally ingesting limewater.

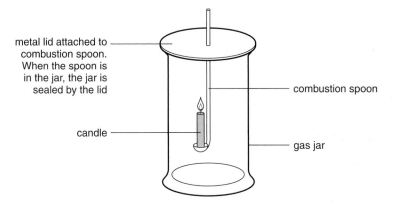

metal lid attached to combustion spoon. When the spoon is in the jar, the jar is sealed by the lid

combustion spoon

candle

gas jar

Figure 5.3 *Estimating the oxygen content of inhaled and exhaled air. The time the candle will burn in each is proportional to the oxygen content*

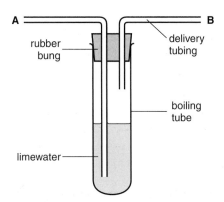

A   B

rubber bung

delivery tubing

boiling tube

limewater

Figure 5.4 *To show that exhaled air contains more carbon dioxide than inhaled air. Inhale through B: air will be sucked through the limewater via A, but the limewater will not turn cloudy, indicating relatively low carbon dioxide levels in inhaled air. Exhale through A: the limewater will turn cloudy indicating the presence of more carbon dioxide in exhaled air*

Research has shown that pupils often think exhaled air contains mainly carbon dioxide and inhaled air contains mainly oxygen (Yip 1998). To avoid such misconceptions, (i) stress that each is only one component; in fact nitrogen is the main component of both inhaled and exhaled air, (ii) give pupils percentage components of inhaled and exhaled air, (iii) avoid using the words 'rich' or 'poor' in reference to a particular gas when discussing the results of the above experiments, and (iv) remind pupils that the 'kiss of life' is successful only because exhaled air contains oxygen.

Assessing pupils' learning

- Ask pupils to write the word equation for respiration for themselves.
- Ask pupils to write up the experiments showing that they understand how the body acquires oxygen and glucose and expels carbon dioxide and water.

KEY STAGE 4 CONCEPTS

How is energy released in cells

At KS4, pupils are likely to need the symbol equation. You can ask more able pupils to write and balance it for themselves. If foundation level students do need the symbol equation, provide it for them.

$$C_6H_{12}O_6 + 6O_2 \rightarrow 6CO_2 + 6H_2O + \text{Energy}$$

Depending on your syllabus, do some of the practicals from KS3 again. Explain that oxygen gets to the cells via the mouth, lungs and blood, and that glucose gets to the cells via the mouth, digestive system and blood. If your muscle cells need more energy, for example during exercise, you would expect your body to increase the glucose and oxygen supply. The following practicals can be used as simple Sc1 investigations of how your body achieves this. They are particularly suitable for lower ability students.

- Breathing rate may increase to gain more oxygen for respiration and expel more carbon dioxide produced by respiration. Ask pupils to measure each others' breathing rate before and after five minutes exercise.
- Pulse rate may increase to pump oxygen and glucose around the body more quickly. Ask pupils to measure their pulse before and after exercise.

At KS4, there are a range of experiments to demonstrate that other living things respire. These test whether living things need oxygen, produce carbon dioxide and release energy. They may be most appropriate for more able pupils. Less able pupils should concentrate on the evidence for respiration in humans from KS3, i.e. that they produce carbon dioxide and water, and need oxygen and glucose.

Do living things need oxygen?

Use the apparatus in Figure 5.5. The alkaline pyrogallol absorbs oxygen. The water acts as a control. Students should predict that the seeds with access to oxygen will germinate, and the seeds without oxygen will not germinate.

S Safety Advice: Alkaline pyrogallol is caustic and you should set this up as a demonstration. For a class practical use iodine solution instead of alkaline pyrogallol and use dead seeds (kill them by boiling for five minutes) in the flask which contains the iodine (don't tell the pupils you have cheated!).

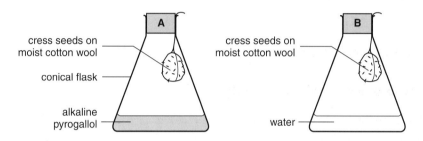

cress seeds on moist cotton wool

conical flask

alkaline pyrogallol

cress seeds on moist cotton wool

water

Figure 5.5 *Apparatus to show that seeds need oxygen to germinate*

Pupils can investigate the rate at which oxygen is consumed by living organisms using the apparatus shown in Figure 5.6. Try using woodlice or maggots on the platform. One mole of oxygen is consumed per mole of carbon dioxide

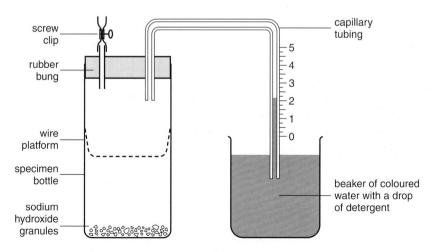

screw clip

rubber bung

wire platform

specimen bottle

sodium hydroxide granules

capillary tubing

beaker of coloured water with a drop of detergent

Figure 5.6 *Apparatus to measure the rate at which oxygen is consumed by living things*

produced. However, the sodium hydroxide absorbs the carbon dioxide, so the water moves up the capillary tube in response to the drop in pressure caused by consumption of oxygen. If you know the dimensions of the capillary tube, you can calculate the volume of oxygen used per unit time: $\pi r^2 h$, where r is the internal radius of the capillary tube and h is the distance travelled by the water.

Do living things produce carbon dioxide?

Use the apparatus in Figure 5.7. The caustic potash removes carbon dioxide from the air entering the chain. The limewater confirms the carbon dioxide has been removed (if carbon dioxide is absent, it will remain clear). Any carbon dioxide produced will be detected in the second flask of limewater. A positive result may need several days to materialise.

- If your result is not convincing, blow into the second flask of limewater.
- If you use a small mammal, use a larger container like a bell jar on a glass plate. Seal the bell jar and glass plate with vaseline. A small mammal should give you a positive result during the lesson.

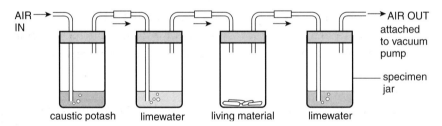

Figure 5.7 *Apparatus to show that living things produce carbon dioxide*

Do living things release energy from food?

The energy in glucose is not totally converted to ATP; some is lost as heat which can be detected (Figure 5.8). Kill the peas in flask 1 by boiling. Wash the live peas with a mild disinfectant to ensure that heat detected is not from respiration by bacteria on the seed surface. If you see no temperature difference in time for your lesson, engineer a temperature difference using warm water in the flask containing live seeds. You could also use a temperature sensor to measure the change in temperature over several days.

Assessing pupils' learning

Pupils should write up the experiments, explaining how they demonstrate the production of carbon dioxide and heat, and the consumption of oxygen.

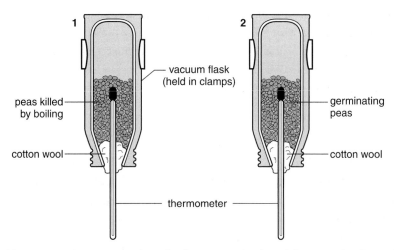

Figure 5.8 *Apparatus to show that heat energy is released from germinating seeds during respiration. Ensure the peas have already started to germinate (soak them in water for two days) before placing them in the flasks*

Respiration without oxygen

There are two forms of anaerobic respiration: lactic and alcoholic. Such respiration is often referred to as fermentation. Alcoholic fermentation occurs in plants and fungi. Lactic fermentation occurs in animals and some bacteria.

Alcoholic fermentation

Place the apparatus in Figure 5.9 in a water bath at 25°C. As you heat the tube, air will escape. Eventually, bubbles of carbon dioxide will pass through the layer of oil, down the delivery tube and into the limewater. This will go cloudy in response.

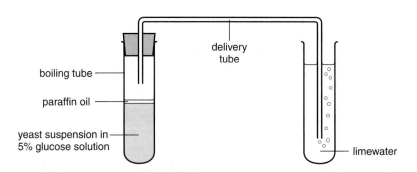

Figure 5.9 *Apparatus to demonstrate that anaerobic respiration (alcoholic fermentation) produces carbon dioxide*

35

Explain that you know the reactants (glucose) and one of the products (carbon dioxide). If you leave the apparatus fermenting until the next lesson, pupils will be able to recognise the smell of the other product – alcohol (ethanol). Hence they can write the word equation and higher level pupils can write and balance the symbol equation:

$$\text{Glucose} \rightarrow \text{Ethanol} + \text{Carbon dioxide} + \text{Energy}$$
$$C_6H_{12}O_6 \rightarrow 2C_2H_5OH + 2CO_2 + \text{Energy}$$

If you have time, introduce pupils to alcoholic fermentation in bread-making. Mix 20 g of strong white (bread-making) flour, 1 g of sugar and 30 ml of yeast suspension. Stir until you have a smooth paste. Pour the mixture into a 250 ml measuring cylinder and place it somewhere warm. The dough will rise up the cylinder because of the formation of bubbles of carbon dioxide. You could investigate the effect of temperature on the speed of respiration by placing the measuring cylinder in water baths of different temperature.

Lactic fermentation

Ask pupils what happens when they exercise hard – their muscles hurt. This is caused by lactic acid acting on pain receptors. Lactic acid is produced by animal cells which cannot get enough oxygen to respire aerobically. This happens if muscles are exercising so much that they need more energy than can be provided given the amount of oxygen available. Excessive lactic acid will eventually inhibit the enzymes involved in respiration, causing the muscle to stop working. You will need to tell pupils the equations for lactic fermentation.

$$\text{Glucose} \rightarrow \text{Lactic acid} + \text{Energy}$$
$$C_6H_{12}O_6 \rightarrow 2C_3H_6O_3 + \text{Energy}$$

Assessing pupils' learning

- Pupils should show written evidence that they understand each type of anaerobic respiration and write up experiments with appropriate conclusions.
- Most able pupils could calculate the energy released from each set of reactions using the bond energies for each compound.

Oxygen debt

Ask pupils to do vigorous exercise. They should measure their pulse rate before, and then every minute afterwards and plot a graph of time against pulse rate. They can do the same for breathing rate, although they will have to measure each other's breathing rate in pairs – it is too difficult to measure your own breathing rate without consciously affecting it. On the graph they can indicate their recovery period (the time taken for the breathing rate and pulse rate to return to normal). Ensure that they exercise for at least five minutes; their muscles should then ache from lactic acid production. Pupils could enter their data into a spreadsheet, and use it to draw the graph.

Ask pupils to explain why we breathe hard during exercise – because our muscles need more energy and are doing more respiration, hence requiring an increased supply of oxygen. Ask them why, once exercise is finished, we need to keep breathing fast. Pupils should realise that it is important to get rid of any lactic acid produced during vigorous exercise. It causes fatigue in muscles, preventing them working properly, and lowers blood pH.

- Explain that the body gets rid of lactic acid by reacting it with oxygen to form carbon dioxide and water. The oxygen required to do this is called the oxygen debt.
- This debt can only be repaid after exercise has finished, and explains why you need to continue breathing fast, and your heart continues to beat fast.
- The recovery period is the time taken to break down the lactic acid.

Assessing pupils' learning

- Ask pupils to write up the exercise experiment, explaining why breathing rate and heart rate remain high after exercise.
- Provide pupils with data on pulse rate and heart beat from a person before, during and after exercise. Pupils should explain the shape of the graph with a series of structured questions.

Reference

Yip, D-Y. (1998) Erroneous ideas about the composition of exhaled air. *School Science Review* **80**:55–62

chapter 6

Nutrition

BACKGROUND

Humans require seven types of food: carbohydrates (for energy), fats (for energy and warmth), proteins (for growth), minerals and vitamins (which have a variety of functions in the body), fibre (to help defaecation) and water (a medium in which chemical reactions can take place). Humans must maintain a balanced diet to receive the correct proportions of nutrients. Before these nutrients can be absorbed into the body, digestion (the break down of food), both mechanical (e.g. with the teeth) and chemical (with enzymes) must occur. When nutrients are released, they are absorbed across the wall of the intestine into the capillaries which surround the gut wall. Different parts of the digestive system are involved in digesting different types of food. Any waste material is ejected from the body by egestion.

Nutrition is included at KS3 and KS4 double and single award. Most pupils already know from KS2 that teeth break down food and that a variety of nutrients are needed to keep healthy. They may also have some background knowledge from home economics. At KS3, pupils must know the components of a balanced diet and their functions, what types of food contain those components, the principles of ingestion, digestion and egestion, and the basic structure of the digestive system. At KS4, they must also know the detailed structure of the digestive system and more detail about digestion.

KEY STAGE 3 CONCEPTS

A balanced diet

Begin by talking about a car. It needs petrol, oil, water and oxygen (for fuel combustion) to function properly. A car is a machine and human bodies are like machines. Ask pupils to think of which nutrients are needed for humans to survive.

Having done this, ask pupils to test for some of these food types: carbohydrates (sugars and starch), fats and proteins. To see the colours produced, you will need foods which are white or can be made into a clear solution or suspension.

Alternatively, ask pupils to test for different nutrients in food first, and then research which other food types are in common supermarket foods from nutritional labels. From this, they should be able to list all the nutrients required by the body.

All foods to be tested must be in solution or suspension. Form a suspension by mashing up food in distilled water and shaking vigorously.

1 Testing for sugar (glucose)

Place 2 ml food solution and 2 ml Benedict's reagent into a test tube. Heat the test tube in a boiling water bath. If sugar is present, a green, yellow, orange and, in the presence of lots of sugar, brick-red precipitate will form. Eye protection must be worn.

2 Testing for starch

Add a few drops of iodine solution to 2 ml of food solution. If starch is present, a blue-black colour will form. Eye protection must be worn.

3 Testing for protein

Add 2 ml Biuret A reagent (20% sodium hydroxide solution) to 2 ml of food solution and shake. Add a few drops of Biuret B reagent (1% copper sulphate solution). If protein is present, a purple colour will form. Eye protection must be worn. If sodium hydroxide solution or copper sulphate solution make contact with the eye, irrigate for 10 minutes and seek urgent medical attention.

4 Testing for fat

Add 2 ml ethanol to a small amount of ground-up food. If testing an oil, simply add the ethanol direct. Shake the mixture well. After allowing to settle, pour off the ethanol into 2 ml of water. If fat is present, a white emulsion will form. If ethanol comes into contact with the eye, irrigate for 10 minutes and seek medical attention. Ethanol is highly flammable; extinguish naked flames before use.

Pupils should know the function of each of the nutrients in the human body. Your department may have a CD-ROM which investigates the effects of diet on health. Take note that most adults think protein is used for energy rather than growth; pupils may have similar misconceptions.

Assessing pupils' learning

- More able pupils could research the functions of specific minerals or vitamins from textbooks or a CD-ROM.
- Less able pupils could match up the nutrient with its function in a cloze-style exercise.

Research has shown that younger pupils often think vitamins are found only in vitamin tablets! To overcome this, pupils could test different foods for vitamin C content. Mash up the food and add water to make a suspension. Add a few drops of blue DCPIP solution (2,6-dichlorophenolindophenol). It turns colourless if vitamin C is present.

Alternatively, you could tell them about Captain Cook feeding his crew with fruit and vegetables (specifically limes) on his explorations around the world: hence providing them with vitamin C to prevent scurvy. Pupils could produce a short play based on the story. This is the origin of the nick-name 'Limeys'.

All pupils must know common sources of nutrients. Ask them to examine packaging at home and make lists of 10 different foods which are rich in each of the different types of nutrient. Collate this information in class and some patterns should result (e.g. rice and pasta are rich in carbohydrates, meat and fish are rich in protein). Pupils could enter data for each food into a spreadsheet and produce a bar graph or pie chart, showing the proportion of different nutrients in each. Other activities include:

- Pupils could survey eating habits within school and produce a report advising the school's kitchens on how to provide a balanced diet. Pupils could use a word processor to design the questionnaire and write the report.
- More able pupils could investigate the prevalence of additives in foods, and research their functions and health implications. Some supermarkets produce useful information leaflets.

Assessing pupils' learning

Ask pupils to

- design a balanced diet for a day.
- analyse menus, identifying in which food each type of nutrient can be found.

Use some of the activities described above to briefly remind pupils about a balanced diet.

KEY STAGE 3 CONCEPTS

Food processing

Begin by defining the terms ingestion (taking in food), digestion (breaking down food into soluble molecules), absorption (taking soluble food molecules from the gut into the bloodstream) and egestion (ejecting waste material from the body via the anus).

Digestion

Most pupils will know that food is broken down. Such breakdown occurs in two stages.

1 Mechanical digestion involves (i) the teeth chopping, grinding and slicing food and (ii) the muscular action of the intestine and stomach churning and pushing the food along.

2 Chemical digestion involves chemical tools called enzymes which break large molecules into small molecules. Pupils should be familiar with an example of enzyme action.

- Ask pupils to test a 2 ml sample of 1% starch solution for starch and sugar as detailed earlier.
- Take another 2 ml of starch solution, add 2 ml 2% amylase solution, leave for 20 minutes, and then test for starch and sugar again.
- The starch will have been broken down into its constituent sugar molecules. Your department may have other favourite practicals which it uses.

S **Safety Advice:** Ensure pupils are wearing safety glasses.

Absorption

Digestion of food is important to produce soluble molecules which are small enough to be absorbed across the gut wall into the bloodstream where they can be carried to the cells which need them. Use a sieve to represent the membrane in the intestine and a digestive biscuit as food. You can mechanically digest it by breaking it into quarters (representing chopping and slicing with teeth, or churning with stomach muscles). Then do some chemical digestion: add water to simulate addition of enzymes and mash the biscuit up. It will then squeeze through the holes, demonstrating the need for digestion.

The apparatus in Figure 6.1 can be used to reinforce the importance of breaking food down to small molecules. Tell pupils that the tubing has tiny holes in and represents the gut membrane. The large starch molecules will not diffuse out of the tubing into the water (representing the blood) whereas the small glucose molecules will. You can test the water for starch and glucose as detailed earlier.

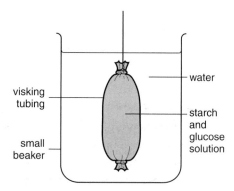

Figure 6.1 *Apparatus to show that only small molecules can be absorbed from the small intestine into the blood. The mixture of glucose and starch represents digested (small) and undigested (large) food molecules, respectively. The water represents the blood into which digested food is absorbed. The visking tubing represents the membrane of the small intestine*

For less able pupils, make a model of the visking tubing to help explain what is happening (Smith & Lock 1992; Figure 4.1). Put large buttons into the bottle to represent starch and small beads to represent glucose. Shake the bottle and the glucose will escape while the starch remains trapped.

Egestion
Younger pupils will attribute the main function of egestion as making room for more food. Explain that the purpose of egestion is the removal of undigestible material from the body through the anus. Some pupils will try to call this excretion – this is wrong. The term excretion applies to the removal of waste which is actually made by the body. In fact, the body adds some excretory waste to the indigestible food, so defaecation is actually egestion *and* excretion.

KEY STAGE 4 CONCEPTS

Food processing

Pupils should be familiar with the above; repeat the practicals to reinforce the ideas. They should also know the mechanism of enzyme action. You will need to explain that enzymes are catalysts – i.e. they speed up reactions without being changed themselves. Research has shown that many KS3 pupils think enzymes are made of cells; this idea may perpetuate into KS4. They are actually large protein molecules which are folded into specific shapes. Somewhere on the enzyme molecule will be an active site. The shape of the active site determines whether it can bind a particular substrate. Different enzymes have different-shaped active sites to which only specific food molecules can bind.

Although you can explain this so-called 'lock and key' hypothesis using the board (Figure 6.2), pupils sometimes fail to understand the three-dimensional nature of what is happening. Use sponges to help demonstrate this (Lester & Lock 1998) – cut active sites out of sponges to fit sponge substrate molecules.

Explain that because enzyme-mediated reactions, just like any other, depend on enzyme and substrate molecules colliding, an increase in temperature increases the speed at which molecules move and therefore increases the rate of reaction.

You can role play this in the sports hall. Use half the pupils as enzyme molecules and half as substrate molecules. Pupils should walk in straight lines, and when they collide with something, bounce off at the angle of incidence. Let them walk for two minutes, during which time each 'enzyme' pupil should count their number of collisions with substrate molecules, and vice-versa. Do the same role play for two minutes but with pupils walking fast – the number of enzyme–substrate collisions should increase.

Because enzymes are proteins, their shape can be changed by increases in temperature and changes in pH. Both affect the bonds holding the protein in its specific shape. If the active site loses its shape (the enzyme denatures), it can no longer bind the substrate molecule. This slows down the rate of reaction. Return to the sponge model and show how easy it is to change the shape of the active site. Each enzyme has an optimum temperature at which the enzyme and substrate

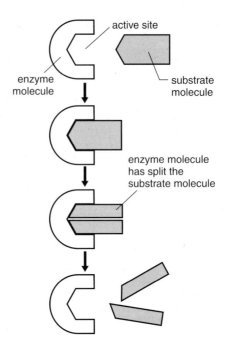

Figure 6.2 *The action of an enzyme on its substrate. The substrate must fit into the active site of the enzyme before it can be digested*

molecules are moving as fast as possible (and hence colliding most frequently) without the active site having changed shape. Their optimum pH is that at which the active site is the best shape to bind the substrate.

The effect of temperature and pH on enzyme action can be used as an Sc1 investigation. The action of trypsin on milk protein gives an easily recordable result. The concentration of milk and enzyme you use will depend on your source of milk (powdered milk tends to work well). Trypsin digests the white protein particles in the milk to amino acids, leaving an almost clear solution (some fat particles remain in suspension in the milk). If pupils carry out a reaction at room temperature and leave it for 30 minutes, they can be fairly sure that all the protein has digested, and will know how clear the milk–trypsin solution should be after complete digestion. They can then set up the same reaction at different temperatures or pH, and time how long each solution takes to reach complete digestion.

S Safety Advice: All pupils should wear safety glasses when using trypsin, acids and alkalis.

The relative concentration of enzyme and substrate molecules also affects the rate of reaction. Using the same role play as before, try varying the numbers of pupils with the roles of enzyme and substrate. You can measure the rate of reaction by counting the total number of enzyme–substrate collisions in a given time. Again, you can modify the trypsin experiment to investigate the effect of enzyme and substrate concentration on rate of reaction.

More able pupils must know that the liver produces bile salts which it secretes into the duodenum. The function of bile is to lower the surface tension of fat droplets to break them up (emulsification) and give a greater surface area for enzyme action. The action of bile is not called chemical digestion because no enzymes are involved. You can demonstrate the action of bile: pour 4 ml cooking oil into two boiling tubes, each containing 20 ml water. Add detergent (represents bile salts) to one, and shake both. The tube with detergent will form a longer-lasting emulsion with smaller fat droplets. Stress that the bile helps to break the fat droplets up and does not affect the shape of any enzymes.

Assessing pupils' learning

- Pupils should write up all experiments, explaining the effect of enzyme action using the lock and key hypothesis and the effect of temperature, pH and concentration on enzyme activity.
- Ask pupils to write about the nature of enzyme action and the effects of denaturation.

KEY STAGE 3 CONCEPTS

The digestive system

Many younger or less able pupils will often suggest that there is only one tube which leads into your body from your mouth. Stress there are two: one into the digestive system (the gullet or oesophagus) and one into the breathing system (the trachea). One provides the food for respiration and one the oxygen.

Discuss the pathway of food from the mouth to the anus. If your department has a model human, use this to aid discussion. Give pupils a diagram of the digestive system to label. They need to know what happens in each part of the system. They could research the functions using textbooks or CD-ROMs, and word process the information obtained in a report or poster.

If you are short of time, ask pupils to match up functions with the correct organ. Stress which of the processes of ingestion, mechanical digestion, absorption and egestion are taking place in each area. Lock et al. (1998) describe some games to help learn the structure and function of the gut.

Pay particular attention to the adaptations of the small intestine for absorbing food molecules. Your department may have pre-prepared slides of a section through the small intestine. Absorption will increase if you can increase the surface area available. Pupils often find this concept difficult to understand. Use a 30 cm piece of drainpipe to represent the small intestine.

- Roll a piece of A4 paper lengthways to represent the membrane on the inside surface of the small intestine, and insert into the drainpipe. Roll a piece of A3 paper lengthways and keep for later.
- Tell pupils that molecules can diffuse through the paper to the space between the paper and the pipe, which represents the blood. That is, molecules are absorbed through the membrane.
- Ask them how we can increase the surface area available for absorption: many will suggest increasing the length. You can remind pupils of the fact that the small and large intestine are very long and coiled in the body for this reason.
- Also suggest to them that if you can increase the amount of membrane without increasing the length of the tube, you will increase absorption even more. Take the rolled A3 paper and ask pupils how it will fit in the 30 cm drain pipe. Most will suggest squashing it: do so, and place it into the tube. When you take it out, you can show them that the paper is highly folded, hence increasing surface area for absorption. This is exactly what happens in the small intestine: the folds are called villi, and they themselves have folds on their surface called microvilli.

Assessing pupils' learning

- Ask questions about the functions of each part of the digestive system.
- Pupils could invent an artificial organism, which is perfectly adapted to obtaining and processing food.
- Pupils should explain why the drainpipe demonstrates how the small intestine is adapted for absorption.

KEY STAGE 4 CONCEPTS

The digestive system

Pupils should be familiar with all the above. More detail will be required than at KS3, the specifics of which will depend upon your syllabus. In general, the types of molecule digested in each part of the digestive system and, for higher level students, the enzymes which carry this out will be needed. Pupils should also know the functions of the pancreas (secrete digestive enzymes into the small intestine) and the liver (receive the food absorbed in the small intestine via the hepatic portal vein and process it, releasing the correct food molecules for transport in the blood around the body).

Higher level pupils must understand peristalsis, the muscular action involved in moving food along the gut. This is a difficult concept: demonstrate it using rubber tubing. Put marks on the outside of the tube every 1 cm. You must use it to demonstrate alternate contractions of the circular and longitudinal muscles in the gut wall. Put a marble into the tube about 10 cm from the end. Stretch the tube (longitudinal muscles relaxed) and squeeze the tube (circular muscles contracted) just behind the marble. Stop stretching (longitudinal muscles contracted) and squeezing (circular muscles relaxed). The marble will have moved along the tube.

You must focus again on absorption in the small intestine. Absorption occurs by diffusion (see Chapter 4). You may need to define diffusion; stress that particles do not make decisions to go into the blood. They simply bounce backwards and forwards from the gut into the blood and they could bounce backwards from the blood into the gut. The good blood supply in the capillaries in the villi ensures this does not happen, because food molecules are carried away before they can bounce backwards. This maintains the concentration gradient and ensures net movement of food molecules from the gut into the blood.

Assessing pupils' learning

In addition to the activities in Key Stage 3 above:

- pupils should rearrange diagrams of contractions of the circular and longitudinal muscles during peristalsis.
- pupils could complete a table showing what is digested where, and with which enzyme.
- more able pupils could complete an independent project on the function of the digestive system.

References

Lester, A. & Lock, R. (1998) Sponges as visual aids – bath time fun for biologists? *Journal of Biological Education* 32:87–89

Lock, R., Bushell, L., Dunn, S., Howden, C., Hughes, P. & Lisle, L. (1998) Digestion aids? Games with guts. *School Science Review* 80:88–93

Smith, G. & Lock, R. (1992) A lemonade bottle model of visking tubing – a model system of a model system? *Journal of Biological Education* 26:8–9

chapter 7

Breathing

BACKGROUND

Breathing is the mechanism by which oxygen is taken into the body for
respiration, and by which carbon dioxide, a waste product of respiration, is
removed from the body. Because an exchange of gases occurs during breathing,
it is also called gaseous exchange. To facilitate gaseous exchange, the lungs
comprise millions of tiny, moist sacs (alveoli), which are surrounded by
capillaries (tiny blood vessels). Oxygen is absorbed into the bloodstream across
the walls of these sacs. Carbon dioxide is removed from the blood into these sacs.
Inflation and deflation of the lungs is achieved by two sets of muscles: the
intercostal muscles (between the ribs) and the diaphragm muscle (a sheet which
divides the thorax from the abdomen).

Many textbooks may still refer to breathing as respiration. This is wrong.
Breathing used to be referred to as external respiration, and the release of energy
from cells, internal or cellular respiration. Discourage this distinction: pupils
must refer to breathing as gaseous exchange. Pupils at KS4 seem never to shake
this misconception and lose marks as a result at GCSE.

Pupils may know from KS2 and their own general knowledge that the lungs
are the location at which oxygen enters the body and carbon dioxide leaves. At
KS3, pupils must understand how the structure of the lungs facilitates gaseous
exchange and how smoking can disrupt lung structure. At KS4 double award,
pupils must understand the structure of the thorax and the muscular actions
which cause the process of breathing to take place. Breathing is not included at
KS4 single award.

KEY STAGE 3 CONCEPTS

Why breathe?

Ask pupils why we breathe (to obtain oxygen for respiration and expel carbon dioxide produced by respiration). If you did not use the practicals described in Chapter 5 for identifying the components of inhaled and exhaled air, do so now. Pupils should conclude that exhaled air contains less oxygen, more carbon dioxide and more water vapour than inhaled air. Many pupils think that you breathe out just carbon dioxide. Work hard to destroy this idea: provide pupils with the constituents of inhaled air and exhaled air. Ensure that pupils can define breathing (the movement of air into and out of the lungs), inhalation (taking air into the lungs) and exhalation (pushing air out of the lungs).

Assessing pupils' learning

- Pupils can write up the practicals from Chapter 5.
- Pupils could write in the first person as if they are a molecule of oxygen or carbon dioxide and describe their path around the body and in and out of the lungs.

KEY STAGE 4 CONCEPTS

Remind pupils of the above, and repeat the practicals in Chapter 5. You could also set up a circus of different organisms (e.g. earthworm, amoeba, plant, insect, fish, bird and mammal) and ask pupils to research how these species gain oxygen for respiration. Pupils should realise that there is a broad range of gaseous exchange systems in different species.

KEY STAGE 3 CONCEPTS

The structure of the thorax and the mechanism of breathing

The thorax is the chest region of the body. Pupils should be familiar with the basic structure; your department may have a life-sized model. Provide a diagram which they can label. You should trace the pathway of air down through the mouth, larynx, trachea, bronchus, bronchioles and into the alveoli. If possible, obtain some pig's lungs to examine. If the oesophagus is attached, point out that it is not part of the breathing system. The trachea is the tube with the rings of cartilage in its wall. You can inflate a lung by cutting open one of the bronchi (it may take some finding) and inserting a plastic tube.

S **Safety Advice:** Use pig's lungs rather than lungs from a sheep or cow. Wear gloves for dissection and wash hands thoroughly afterwards. Do not allow pupils to inflate lungs. Sudden deflation of the lung may propel blood back into your mouth.

To help make pupils realise what happens to their ribs when they inhale and exhale, ask them to measure their lung capacities. Your department may have lung capacity bags with volumetric gradations marked. If not, exhale into a large measuring cylinder over water. Pupils can measure their tidal volume (the amount of air normally breathed in and out) and vital capacity (the largest amount of air you can breathe in and out).

S **Safety Advice:**

- Ensure mouthpieces or tubes for collecting exhaled air are disinfected before and after use.
- Ensure that pupils do not hyperventilate; intake of too much oxygen causes severe muscle cramps. Hyperventilation has traditionally been treated by breathing in and out of a paper bag.

Ask pupils what happens to their ribs when they breathe in and out. They are raised during inhalation and lowered during exhalation. You will need to tell pupils that they have muscles (intercostal muscles) between their ribs. When these contract, the rib cage lifts. You will also have to tell them they have a sheet of muscle between their thorax and abdomen (the diaphragm): singers may be familiar with this. The muscles on this sheet contract and the sheet lowers during inhalation. The volume of the thorax therefore increases when you inhale. Use the apparatus in Figure 7.1 to explain what happens to the lungs when the diaphragm is lowered and raised. Pupils can make their own models using a plastic beer glass or bottle. If the beer glass is flexible enough, it may also demonstrate the movement of the rib cage, in and out.

More able students may understand that an increase in the volume of the thorax will lower the air pressure within it, and air will rush in through the mouth from the outside. When the volume of the thorax is reduced during exhalation, the pressure increases and air rushes out of the lungs to the outside. Avoid talking about air pressure with less able pupils.

> ### Assessing pupils' learning
>
> - Pupils should label the structure of the thorax.
> - Ask pupils to compose a flow chart, showing the events occurring during inhalation and exhalation.
> - Ask pupils to compose a second flow chart, showing the structures through which air passes in and out of the lungs.

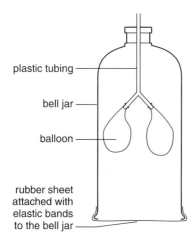

plastic tubing

bell jar

balloon

rubber sheet
attached with
elastic bands
to the bell jar

Figure 7.1 *Model of the thorax. Simulate inhaling by pulling down the rubber sheet (lower the diaphragm). Because the pressure in the bell jar reduces as a result, the balloons (lungs) inflate. Push the rubber sheet up to simulate exhaling. Because the pressure in the bell jar increases, the balloons (lungs) deflate*

51

KEY STAGE 4 CONCEPTS

The structure of the thorax and the mechanism of breathing

Revise work from KS3. At KS4, more detail must be added. Pupils should be familiar with the function of the components of the thorax, including the pleural membranes (produce pleural fluid) and fluid (lubricate movement of the lungs), and the rings of cartilage surrounding the trachea (support the tube during changes in air pressure). You can demonstrate what would happen without cartilage by blowing or sucking through a long, thin flexible plastic tube (you can make one out of a plastic bag). Because faster moving air (within the tube) is at a lower pressure, the higher air pressure (outside the tube) pushes the tube closed, preventing inhalation and exhalation.

Inhaling is described in KS3 Concepts (above). However, it is actually the external intercostal muscles which contract during inhalation and the internal intercostal muscles which relax. During exhalation, the reverse is true. You can demonstrate this using a model of the rib cage (Figure 7.2) (Lock & O'Hara 1996). For more able pupils, explain the flow of air into and out of the lungs as a result of pressure differences caused by the increase or decrease in size of the thorax.

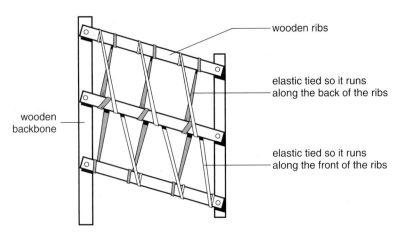

wooden ribs

elastic tied so it runs along the back of the ribs

wooden backbone

elastic tied so it runs along the front of the ribs

Figure 7.2 *Model of the intercostal muscles and the ribs. The thinner elastic represents the internal intercostal muscles. The external intercostal muscles contract (shorten) when the ribs rise. The internal intercostal muscles contract when the ribs lower*

Assessing pupils' learning

In addition to those activities described in KS3 above:

- pupils should explain the results of the plastic tube demonstration, explaining how it shows the need for the rings of cartilage.
- pupils could make their own models of the muscles associated with the rib cage.

KEY STAGE 3 CONCEPTS

Adaptations of the lungs to gaseous exchange

Begin by retracing the pathway of air into the lungs. Before it enters the lungs, the air is cleaned by hairs in the nose (and warmed and moistened). It then passes into the trachea. Lining the trachea is a layer of mucus (sticky slime) and cilia (tiny hairs). The mucus traps dust particles and microbes before they enter the lungs. The hairs push the mucus up to the throat where it is swallowed and any microbes are then destroyed by stomach acid.

Once the air has reached the end of the bronchioles it enters tiny sacs called alveoli (singular: alveolus). These sacs are the site of gaseous exchange (the process where oxygen enters the blood and carbon dioxide leaves it). If you have a set of pig's lungs, when you inflate them, the spongy nature of the lungs should be clear. Suggest to pupils that, like a sponge, the lung contains numerous air sacs (they are individually too small to see). Your department may have prepared slides of alveoli for pupils to examine.

S **Safety Advice:** Be careful of transmission of blood back up the tube when the lung deflates. Do not allow pupils to inflate lungs.

The presence of so many alveoli creates a huge area for gaseous exchange – in an adult, up to the same area as a tennis court. Pupils often find it difficult to understand how surface area increases absorption. Remind them of the adaptations of the small intestine to absorption of food molecules (see Chapter 6), and the explanation using paper and a drain pipe. Tell pupils that making lots of small bags, rather than lots of folds, is simply another way of increasing the area available for absorption.

- Reinforce the importance of surface area using the apparatus in Figure 7.3.
- Less able pupils could make a model of a bronchiole and alveoli using rubber tubing and egg boxes. Use red and blue wires intertwined to demonstrate oxygenated blood arriving at the alveoli and deoxygenated blood leaving.

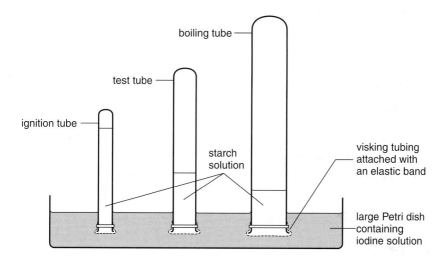

Figure 7.3 *Apparatus to show that importance of a large surface area for absorption. Each tube contains an equal volume of starch solution. The iodine diffuses through the visking tubing into the starch solution, turning it black. The starch in the boiling tube will turn completely black before the starch in the ignition tube because the tube has a larger surface area across which absorption of iodine can occur*

Assessing pupils' learning

- Pupils should produce a flow chart, describing how the air is cleaned when it enters the lungs.
- Pupils could make models of the breathing system as a whole, including the lungs using pipes and sponges.
- Pupils should explain how the lungs are adapted to gaseous exchange.
- Pupils should write up the surface area experiment and relate it to the alveoli in the lungs.

Adaptations of the lungs to gaseous exchange

Pupils should be familiar with all the KS3 work above. Adaptations of the alveoli need to be considered in more detail at KS4. These adaptations apply both to oxygen being absorbed into the blood, and carbon dioxide being expelled from the blood.

If oxygen is to go from being a gas, to being dissolved in the blood, then it must become dissolved in water at some point. The walls of the alveoli are moist, and oxygen dissolves in this moisture. The air sacs also have very thin walls, allowing oxygen to diffuse through into the blood vessel.

If pupils are not familiar with diffusion, you will need to explain it (see Chapter 4). You can also suggest that the excellent blood supply to the alveoli carries oxygen away and allows the maintenance of a concentration gradient whereby oxygen will continue to diffuse from the air sac to the blood. Similarly the removal of carbon dioxide during exhalation maintains the concentration gradient of carbon dioxide from the blood to the air space in the alveolus.

Assessing pupils' learning

Pupils should demonstrate written understanding of the process of diffusion and how the adaptations of the alveoli help absorption of oxygen into the blood and expulsion of carbon dioxide from the blood. This may be best done using annotated diagrams.

Effect of smoking on the lungs

Set up a smoking machine (Figure 7.4) to show pupils the pollutants inhaled during smoking. The cotton wool will discolour, and water vapour from the smoke will condense in the U-tube and appear dirty.

Ask pupils to research the effects of smoking on the lungs. Conclude that smoking reduces the amount of oxygen which can be absorbed into the blood, either because the alveoli break down (emphysema), lung cancer forms (and occupies space previously occupied by alveoli) or because less oxygen can be inhaled (bronchitis).

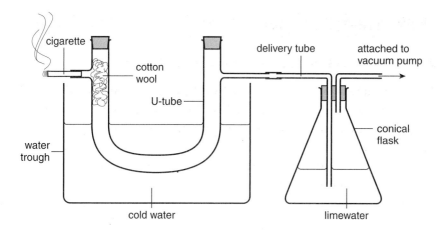

Figure 7.4 *The 'smoking machine'*

Assessing pupils' learning

- Assess pupils' research by asking them to produce leaflets or posters on the symptoms of bronchitis, emphysema and lung cancer.
- As extension work, more able pupils could consider some of the ethical issues connected with smoking. For example, should smokers be entitled to health care?
- Pupils should write up the smoking machine demonstration. They should relate their findings to the effect of smoking on the lungs.

KEY STAGE 4 CONCEPTS

The effects of tobacco on health are considered in Chapter 14.

References

Lock, R. & O'Hara S. (1996) A simple modification to a lung model. *Journal of Biological Education* **30**:240–241

chapter 8

Circulation

BACKGROUND

The circulation has two components: the blood system and the lymphatic system. The lymphatic system is not included at KS3 or 4. The blood system transports dissolved gases such as oxygen and carbon dioxide, and nutrients such as carbohydrates, fats, amino acids and minerals. The human blood system is a so-called double circulation, which means that the blood passes through the heart twice in each complete circulation. The heart pumps blood to the lungs to be replenished with oxygen and to have carbon dioxide removed. When the blood returns to the heart, it is then pumped around the body where oxygen is removed (and used by respiring cells), and carbon dioxide is released into the blood. Food molecules enter the circulation via the hepatic portal vein and the liver. The blood is transported in three types of blood vessel: arteries carry blood away from the heart; veins carry blood towards the heart; and capillaries carry blood between arteries and veins. The different components of the blood have different functions: (i) the red blood cells carry oxygen, (ii) the white blood cells defend the body against disease, (iii) the platelets are involved in sealing cuts, and (iv) the plasma carries dissolved gases, waste and nutrients.

At KS3, pupils should appreciate that blood has several components and understand how the blood acts a transport medium, providing each body cell with the substances needed for respiration (oxygen and glucose) and removing the products of respiration (carbon dioxide and water). Pupils should also realise how exchange of substances happens between capillaries and body cells. Some pupils will say that oxygen is only needed for the heart to pump blood around the body. This is only true to the extent that the cells of the heart require oxygen for respiration, just as all the other cells in the body require oxygen, and are supplied with it by the blood. At KS4 double award, pupils must know the structure of the circulatory system, including the structure of arteries, veins and capillaries, and their location and function. They must also understand the function of each component of the blood. Pupils at KS4 single award must know the components of blood, the function of each, and that blood is a transport medium.

KEY STAGE 3 CONCEPTS

The circulatory system

Remind pupils that we need glucose and oxygen for energy. Introduce the circulatory system as the mechanism by which food and oxygen are taken to the cells from the digestive system and breathing system, respectively. The circulatory system also carries carbon dioxide and waste away from the cells. Discuss with pupils what is required for an effective system: tubes which pass around the body, a liquid to carry substances and a pump.

Reinforce the idea that blood transports the reactants and products of respiration around the body:

- Pupils should wrap an elastic band around their fingers. This should not be too tight.
- With the elastic band in place, and their arm by their side, they should open and close their fingers as many times as possible for 2 minutes.
- Having rested their hand, repeat the experiment with their arm raised above the head.

When your arm is above your head, the blood is being pumped uphill. As a result, it reaches the hand more slowly. It therefore supplies the raw materials for respiration more slowly. Because of this, less energy is released in your muscles, and you can do less exercise with your fingers.

S **Safety Advice:** Warn pupils to slow down when their muscles begin to hurt. They should not put excessive strain on their hands.

Tell pupils there is a double circulation. You could set up a role play to represent this. Construct the pathways of the double circulation with tables and chairs, or draw out paths on the classroom floor. Use different coloured cards to represent oxygen and carbon dioxide and make pupils walk around the circulatory system, gaining and losing oxygen, nutrients and carbon dioxide as they go.

Assessing pupils' learning
- Give pupils a schematic diagram of the double circulation and ask them to colour in red and blue to show the locations in which blood is oxygenated or deoxygenated.

• Ask them to translate their role play onto a paper diagram.
• Ask pupils to write up the elastic band experiment, explaining how it demonstrates the need for the circulatory system.

KEY STAGE 4 CONCEPTS

Pupils should be familiar with all of the above. You may need to remind pupils why oxygen and food are needed (respiration to release energy), and where they are needed (in every body cell).

KEY STAGE 3 CONCEPTS

The heart

Carry out a demonstration dissection to show the structure of the heart. Ensure you have looked at the heart before the lesson; butchers often cut off the main veins and arteries and this makes it more difficult to identify each blood vessel and to show pupils the pathway through which blood flows. Ensure pupils realise that when they look at the heart, its left side is on the pupil's right. If they do not understand, ask them to imagine the heart inside a person lying on their back.

If you cannot get a heart for dissection, try making a model out of sponge (Lester & Lock 1998), or view the dissection on the internet at http://sln2.fi.edu/biosci/preview/heartpreview.html.

In either case, point out that the right ventricle wall is thinner than that of the left ventricle. This is because the left ventricle needs extra muscle to pump blood to the whole body. The right ventricle cannot pump blood at high pressure to the lungs or they may be damaged.

S **Safety Advice:** If you want to do this as a class dissection, check current DfEE regulations and consult your head of department.

If pupils dissect the heart, they should make a drawing, preferably of the cross-section of the heart. If not, they should be provided with diagrams to label. They should indicate the flow of blood through the heart with red and blue arrows, denoting oxygenated and deoxygenated blood, respectively.

To show the importance of the heart in circulation, ask pupils to take their pulse before and after exercise. They should explain their results as follows: (i) exercise demands more energy than resting, (ii) muscle cells must therefore respire more quickly, (iii) food and oxygen must be pumped around the blood more quickly for more respiration to occur, (iv) therefore pulse rate increases.

KEY STAGE 4 CONCEPTS

The heart

Again, begin by examining the heart. Single award pupils do not need to know the internal structure of the heart, although it will be of interest to see or do the dissection. Double award pupils should know the structure of the heart (the atria, the ventricles and the major blood vessels), and should be able to describe the pathway of blood through the heart during the cardiac cycle (Figure 8.1). Pupils should explain whether the atrial or ventricular walls are contracted or relaxed, and which valves are open or closed at each stage. Remember to mention that the left ventricle wall is thicker than the right because it has to pump blood around the whole body.

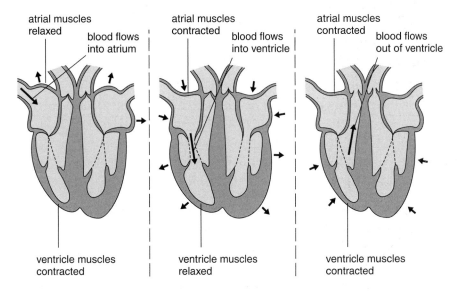

Figure 8.1 *The cardiac cycle*

For more able double award pupils you will need to explain the function of valves. The bicuspid valve is situated between the left atrium and ventricle. The tricuspid valve is situated between the right atrium and ventricle. These valves prevent blood passing up into the atria when the ventricles contract. During dissection, you can highlight the tendons (heart strings) which attach these valves to the wall of the heart: they prevent the valves themselves being turned inside out up into the atria. The semi-lunar valves in the aorta and pulmonary artery prevent blood sucking back into the ventricles when they stop contracting and open again ready to receive blood from the atrium.

Assessing pupils' learning

- Pupils could produce a flow chart to demonstrate blood flow through the heart.
- Pupils should annotate diagrams of the heart at different stages of the cardiac cycle. They should know which muscles are contracted (atrial or ventricular), and which valves are open or closed at each stage.
- Ask pupils to explain why they can feel their pulse.

KEY STAGE 3 CONCEPTS

The blood vessels

You can show pupils veins and capillaries. All pupils can observe veins under their skin: the blue blood vessels visible on the underside of your lower arm. Rub cedarwood oil on your finger, between the nail and first knuckle, and examine it under bright light and low magnification; you can observe the capillaries. Use this as a class demonstration if your school has a microscope camera whose image can be viewed on a television screen.

Your department may have prepared microscope slides of the three types of vessel in cross-section. Ensure you explain how the slide was obtained, i.e. by cutting across the blood vessel and looking at the cut end. More able pupils could draw them. All pupils should be given schematic diagrams. Ensure pupils understand the following points:

- Because the heart pumps blood at a high pressure, the arteries need thick, elastic walls.
- Capillaries have extremely thin, one-celled walls which allow transport of oxygen, nutrients, carbon dioxide and waste into and out of the blood.
- Veins have thinner walls than arteries, but thicker walls than capillaries: they do not have to withstand high pressures, but do not need to allow transport of food, oxygen, carbon dioxide or waste into or out of neighbouring cells.
- Veins have valves: these stop blood flowing backwards when it is passing back to the heart at low pressure.

Pupils can make models of the blood vessels to reinforce these points. You will need a piece of rolled up paper to represent the capillary wall (pupils can draw cells on the paper to show it is one cell thick) and cardboard tubes of different diameter to represent the lining (endothelium) of the artery or vein. Roll different amounts of foam or cotton wool around the tubes to represent the walls. You can even make paper semi-circular valves for the veins, fixed on with a Sellotape hinge.

Assessing pupils' learning

- Pupils could write, in the first person, about the journey of a blood cell around the body.
- Pupils could research the differences between the types of blood vessels, and make a table to compare them (wall thickness, direction of blood flow and type of blood carried: oxygenated or deoxygenated).
- More able pupils should draw blood vessels from the microscope slide. Less able pupils could be provided with diagrams to label and annotate.
- More able pupils should explain how the structure of the vessels relates to their function; e.g. arteries have thick, elastic walls to withstand the high blood pressure from the heart.
- Ask pupils to identify the only artery which carries deoxygenated blood (pulmonary artery to the lungs), and the only vein which carries oxygenated blood (pulmonary vein from the lungs).

KEY STAGE 4 CONCEPTS

The blood vessels

Remind pupils of most of the above by repeating some or all of the material depending on your syllabus. Pupils should relate the structure of arteries and veins to their function. More able pupils could do this by research. Less able pupils may need some form of cloze exercise to do this successfully. Mention arterioles (the link between arteries and capillaries) and venules (the link between veins and capillaries).

Pupils must realise how capillary structure relates to function. Stress that materials are transferred between the blood system and the body's tissues via the capillaries. The body's cells are all bathed in 'tissue fluid', which is plasma that has leaked out of the capillaries. Oxygen and nutrients can diffuse (see Chapter 4) out of the capillary through its thin wall. Some mass flow of oxygen and nutrients occurs when plasma leaks out through the gaps in the wall. They then diffuse through the tissue fluid to the cells. Carbon dioxide and waste undergo this journey in reverse.

For less able pupils, set up a role play to demonstrate this, using chairs as the wall of the capillary and have pupils squeezing through the gaps in the chairs to simulate exchange of materials. They could carry cards to represent oxygen and carbon dioxide.

Assessing pupils' learning

- Pupils should explain the adaptations of the veins, arteries and capillaries to their function.
- Pupils should write about the processes involved in exchange of materials between capillaries and body cells.

KEY STAGE 3 CONCEPTS

The components of blood

Pupils should be aware that the liquid part of the blood is plasma, in which the cells float, and in which gases, nutrients and waste are dissolved. Research has shown that it is difficult to convince younger pupils that blood is not just a liquid. Try the following:

- Your department is likely to have prepared slides of blood, showing red and white cells which pupils could draw. Provide less able pupils with prepared diagrams. DfEE regulations prohibit you making slides of your own blood.
- Ask pupils to make polystyrene models of the red and white blood cells, and of platelets. Place them in a beaker of cold tea (represents the plasma). Because they are polystyrene, the models will float in the beaker.
- Centrifuge some pig's blood and point out the difference between the red pellet and the orange plasma.

The adaptations of red blood cells lend themselves to three neat practicals:

1 Red blood cells are filled with a pigment called haemoglobin which binds to and carries oxygen. The red blood cells have no nucleus so they have more space for haemoglobin. If you can obtain some fresh pig's blood from a slaughter house, you can demonstrate the binding of oxygen to haemoglobin. As soon as you get it, add 5 cm³ of 0.1% sodium oxalate per litre of blood to stop it clotting. Place equal amounts of blood in three flasks and bubble oxygen through one, carbon dioxide through another and leave the other as a control. The blood in the oxygen flask will turn bright red, showing that haemoglobin will bind oxygen in a region of high oxygen concentration. The

blood in the carbon dioxide flask will turn blue, showing oxygen is released from haemoglobin in regions of low oxygen concentration.

S **Safety Advice:** Carry out this experiment as a demonstration. Avoid using blood from sheep or cows.

2 Explain how the biconcave shape of red blood cells gives them more surface area for absorption of oxygen. Pupils can make a model of a red blood cell using a sponge (Lester & Lock 1998). If they cover the model with small pieces of paper, and then measure their area, they can compare the surface area of a biconcave and normal sponge.

3 When red blood cells pass through capillaries, they are frequently wider than the capillaries themselves, and they pass along the capillary in single file. This slows them down to allow more exchange of materials to occur. Demonstrate this by pushing a sponge through a large measuring cylinder. Although it can bend (as can the red blood cell), it is still slowed down.

White blood cells come in two types: phagocytes (eating cells) and lymphocytes. Phagocytes squeeze between gaps in capillary walls, wrap themselves around bacteria and engulf them. Lymphocytes produce antibodies.

Platelets are tiny fragments of cells which play a role in clotting blood, and forming scabs. The time at which you teach the defence mechanisms of the blood will depend on your scheme of work. Details are included in Chapter 14.

63

Assessing pupils' learning

- Pupils should write up the experiments and demonstrations described above, showing that they understand the adaptations of the red blood cells.
- Split your class into groups and let each group research a booklet or presentation on one component of the blood. Younger or less able pupils could produce a wall display of the different types of cell, with annotations explaining the function of each.
- Pupils could write a job application for the position of oxygen transporter, identifying the key adaptations of the red blood cell.

KEY STAGE 4 CONCEPTS

The components of blood

At foundation level, repeating KS3 work will suffice. At higher level, pupils need more detail. Red blood cells (erythrocytes) carry oxygen. When oxygen combines with haemoglobin in the red blood cell, it forms oxyhaemoglobin. This reaction is in equilibrium, and whether the oxygen stays combined depends on the relative concentrations of oxygen around the red blood cell.

$$\text{Oxygen} + \text{Haemoglobin} \rightleftharpoons \text{Oxyhaemoglobin.}$$

If pupils have discussed equilibria in chemistry, you can ask them to predict under what conditions the equilibrium will move to the left and to the right. They should suggest that oxyhaemoglobin will be formed where there is a large supply of oxygen (e.g. in the lungs), and oxygen will be liberated where oxygen is lacking or being used up (e.g. body cells which are respiring). Use the blood experiment from KS3 to reinforce this.

Many exam questions concentrate on the adaptations of the red blood cells including:

- presence of *haemoglobin* (which binds oxygen)
- they have *no nucleus*, giving them a large space for haemoglobin, allowing them to carry as much oxygen as possible
- *thin and permeable* (to oxygen) *membrane*
- *flexibility* which allows them to pass through the smallest capillaries,
- *biconcave shape* which gives more surface area for absorption of oxygen. To help understanding, repeat the sponge explanation from KS3.

Mention the plasma as being the main transport medium. It carries dissolved substances (apart from oxygen), such as carbon dioxide from respiration, and urea from the liver. All pupils should predict where these compounds leave the body: carbon dioxide in the lungs, and urea in the kidneys. Because water has a high specific heat capacity, the plasma is also an excellent medium through which to dissipate heat around the body.

Assessing pupils' learning

- Use some of the activities described in KS3 above.
- Ask pupils to make up a table of everything carried by the blood: where it is carried from, where it is carried to, and by which component.

References

Lester, A. & Lock, R. (1998) Sponges as visual aids – bath time fun for biologists? *Journal of Biological Education* 32:87–89

chapter 9

Movement

BACKGROUND

To move, humans need muscles and bones. Bones are connected by joints which can flex (bend) and extend (straighten). Muscles pull on the bones and make them move in relation to other bones. Without bones, muscles would have nothing to pull on. When muscles pull, they contract (reduce in length and become thicker). When they are not pulling, they relax (increase in length and become thinner). Muscles can only pull, they cannot push. To move a bone in two opposite directions, a pair of so-called antagonistic muscles are needed.

At KS3, pupils must understand the structure of the skeleton and the role of joints and muscles in movement. The skeleton and movement is not included at KS4 double or single award. Pupils will have a broad knowledge of movement from KS2 and are likely to have studied the skeleton before. They should also know that muscles make bones move, and some will know that muscles contract and relax.

Introduction

Begin by asking pupils why we need to move. Possible ideas include (i) to find food, (ii) to find a mate, (iii) to find shelter, and (iv) to avoid predators. Bring these points out in discussion and make a written record.

KEY STAGE 3 CONCEPTS

The skeleton

Most biology departments will have a skeleton to use for class discussion. Pupils should realise that we need lots of different bones to allow movement. If all our bones fused together, we would not be able to move. Test how familiar pupils are with the bones in the skeleton, and begin to use scientific names: e.g. funny bone becomes humerus, back-bone becomes vertebral column.

Some pupils will try to call the vertebral column the spinal cord. This is wrong. The spinal cord is the nerve cord which runs through the spine, not the bones themselves. Also point out those parts of the body which are made up of lots of bones; for example there is no 'hand-bone', the hand is made up of lots of bones. Similarly, there is not really a back-bone – this is made up of lots of small bones called vertebrae (singular: vertebra). Ask pupils why there are so many bones in these structures: they allow more complex and dextrous movement.

Pupils will always ask you to identify which is the ulna and radius in the forearm, and which the tibia and fibula in the lower leg. Check before the lesson, and commit it to memory. It is very easy to forget which is which!

Assessing pupils' learning

- Pupils should label a diagram of the skeleton.
- Less able pupils could cut out drawings of the major bones and rearrange them into the correct position to make up a skeleton.
- Ask pupils to imagine that their bones are fused together and that they must invent a new method which could have evolved in humans to allow them to move.

Joints

To move bones in relation to each other, we must have joints. A joint is a junction between two bones. If there was no movable joint between bones, movement would be impossible. Pupils could research the nature of the different types of joint from textbooks or CD-ROMs: namely, ball and socket (e.g. shoulder, hip), hinge (e.g. elbow, knee), sliding (e.g. spine, fingers) and fused (e.g. bones in the skull).

Ensure that pupils have diagrams of ball and socket joints and hinge joints to label. They should identify each of the four types of joint on their own bodies, and identify the range of movement that each allows. They should also know the function of the different components of the joints. To understand why each component is there may require class discussion.

- The muscle must be there to pull on the bones.
- The tendons connect the muscle to the bone. If you can get hold of chicken feet, and dissect away the loose flesh, you can show the tendons. If you pull the tendons, you can make the bones move. Pupils can also tense their wrists; they should be able to feel tendons sticking out from under their skin. If they

squeeze their wrist with the opposite hand, this will pull on the tendons and their fingers will bend.

- The ligament connects bone to bone to prevent the joint coming loose.
- Explain the function of the synovial fluid by talking about a car. The parts in a car's engine only move in relation to each other because they are lubricated by oil. The synovial fluid acts like an oil.
- The cartilage cushions the ends of the bones against each other.

Ball and socket, hinge and sliding joints are all examples of synovial joints (they have synovial fluid). The fused joint is not a synovial joint and allows no movement. The easiest fused joint to see on a skeleton is between the plates in the skull. Run your hand across your temple, you can probably feel a dent which you can tell pupils is a joint between two plates. These plates used to move freely in relation to each other during birth and infancy. Ask pupils why this is advantageous: it allows the baby's head to pass through the birth canal more easily, and allows growth. The plates fuse together as the baby gets older to provide a robust, shock-absorbing container for the brain.

Assessing pupils' learning

- Pupils should show written understanding of the functions of the different parts of the joints, knowledge of the different types of joint, and their range of movement.
- Pupils could research joints from textbooks and CD-ROMs and produce a booklet about them, or could answer questions based on the above information.

Muscles

Begin by allowing pupils to investigate their own muscle action. This will allow you to assess pupils' prior understanding. Give them a range of movements to carry out, and ask them to identify which muscle is causing that movement. The muscle which pulls on the bone (contracts) will not always be the muscle they feel: some will identify muscles which are being stretched as responsible for moving a bone.

Once they have completed the practical, discuss the elbow joint to help clarify how muscles pull on bones. Use an overhead transparency or a model arm to help your explanation. Research suggests that using models and having pupils feel the action of their own muscles is important to help understanding.

Tell pupils that muscles can only pull, they cannot push. If a muscle is pulling, it is contracted (short and thick). If a muscle is not pulling, it is relaxed (long and thin). Tell the class to hold out their arms and bend their elbows. They should indicate that the muscle on the front of the upper arm is contracting and therefore pulling the forearm upwards. You can introduce the name of this muscle: the biceps. Most pupils will realise that the muscle at the back of the upper arm, the triceps, pulls the lower arm back down again. Less able pupils could make a model of the arm (Figure 9.1).

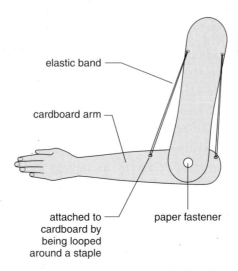

elastic band

cardboard arm

attached to cardboard by being looped around a staple

paper fastener

Figure 9.1 *A model arm. Cut the arm shapes out of stiff cardboard. Ensure your elastic bands are the correct size: the 'biceps' band should bulge when the elbow is flexed; the 'triceps' band should bulge when the elbow is straightened*

Because the biceps bends (flexes) the elbow, it is known as the flexor muscle. Because the triceps muscle extends the elbow, it is the extensor muscle. Explain that muscles which pull the same bone in opposite directions are an antagonistic pair.

Assessing pupils' learning

- Check pupils have identified the muscle which was pulling (contracted), which was the extensor and which the flexor during their movements.
- Give pupils descriptions of other movements and ask them to indicate the positions of the flexor and extensor muscles.
- Give pupils a diagram of the elbow joint both flexed and extended. Ask them to label the diagrams indicating which muscle is contracted and relaxed in each case.
- Ask pupils which muscles are contracting during push-ups and pull-ups.
- Ask pupils to explain why musculature would be more complex at a ball and socket joint which moves in all directions, not just backwards and forwards.
- Ask pupils to give the location of different antagonistic pairs of muscles around the body.

chapter **10**

Reproduction

BACKGROUND

Reproduction allows individuals to pass their genes on to the next generation. Without reproduction, a family line (genes) would die out and a species would die out. Sexual reproduction produces individuals which are different from each of their parents. Asexual reproduction produces offspring which are identical to their parent. Humans reproduce sexually. In humans, both males and females have gonads (testes and ovaries, respectively) which produce the gametes (sperm cells and egg cells (ova)). Because fertilisation and development of the baby occur inside the female, the male needs an organ to introduce sperm into the female; this is the penis. The female also needs to provide a site where the baby can develop; this is the uterus (womb). Once the sperm fertilises the egg, the zygote produced divides repeatedly and the resultant embryo implants itself into the uterus wall where it draws food and oxygen from the mother through the placenta. It also expels waste and carbon dioxide through the placenta.

Human reproduction is only included at KS3. Hormonal control of reproductive physiology is included at KS4 double and single award (see Chapter 12). Pupils from KS1 and 2 should be aware that humans produce babies, and should understand the roles of men and women in supplying eggs and sperm during intercourse. However, the depth with which primary schools teach sex education may vary considerably, with some almost refusing to teach it altogether. Pupils will arrive at KS3 with a wide variety of knowledge, acquired from science lessons, personal and social education, the television and from their parents, siblings and friends. Because of these disparate sources of knowledge, it is likely that many of them will hold quite serious misconceptions.

Before you begin teaching reproduction, read your school's sex education policy and speak to the PSE co-ordinator. Henderson (1994) offers valuable advice about teaching reproduction. However, it is important to be aware that many pupils will be apprehensive about the topic, will feel vulnerable to ridicule, and will be worried about asking/answering questions. Because most pupils are fairly shy, it may be difficult to assess their misconceptions.

Try to create a supportive learning environment where pupils feel able to talk about their understanding. Henderson (1994) suggests asking pupils to define the rules by which the class will operate: for example, confidentiality within the group, no questions about an individual's sexual experience etc. Once pupils have defined the ground rules, you must police them, i.e. you must define the sanctions for breaking those rules. If pupils do not make the rules, put yourself in the position of the least confident child and define rules appropriately. Pupils will feel more confident about contributing if they know they can do so without fear of being wrong or ridiculed.

Pupils will also be more confident if you show that use of words, such as penis or vagina, causes you no embarrassment. This may be difficult if you have not taught sex education before. In this respect, choose teaching methods appropriate for you. For example, if you are not happy to talk about sexual intercourse with a group of 30 year 9 pupils, then show them a video about which they can answer questions. If you use class discussion, bring pupils around your desk; it makes discussion more productive and avoids giggles.

KEY STAGE 3 CONCEPTS

Asexual and sexual reproduction

An easy way to take the wind out of 'silly' children's sails is to start by asking why organisms need to reproduce. The answers pupils give will vary depending on their ability. A possible list includes: (i) to replace individuals that die of old age, (ii) to make sure a species does not die out, (iii) to pass genes onto the next generation.

Next, compare the two methods used by organisms to reproduce: asexual and sexual. Many plants reproduce asexually: they simply grow a new individual from the old one (see Chapter 16). You can show pupils a spider plant with runners (long shoots) and 'baby' plants attached. Discuss as a class the way in which this type of reproduction differs from the way in which humans are produced. For example, (i) number of parents, (ii) formation of gametes, (iii) speed of reproduction, (iv) number of genes from each parent possessed by the offspring, and (v) whether offspring are identical or not identical to the parent.

- Ask pupils to design an artificial organism with a mechanism for reproducing sexually, and an organism with a mechanism for reproducing asexually.
- Pupils should complete a table of the similarities and differences between sexual and asexual reproduction.
- Pupils could research the range of species using asexual reproduction.

KEY STAGE 4 CONCEPTS

Aspects of asexual and sexual reproduction covered at KS4 are discussed in Chapters 19 and 20.

KEY STAGE 3 CONCEPTS

Gametes and fertilisation

Humans reproduce sexually. Because of this, they produce gametes which fuse together (in fertilisation) to form a zygote (one cell). This is the beginning of a new person. You can ask pupils how the egg and sperm are adapted to their jobs in fertilisation. For instance, the sperm cell has a tail with which to swim to the egg, and the egg cell is large to store food for the developing embryo. At the front part of the head of the sperm is a region called the acrosome that contains digestive enzymes which it uses to break into the egg. Most pupils will be able to predict the function of those enzymes; they know that enzymes are used to cut biological molecules.

Assessing pupils' learning

Pupils could:

- draw the sperm and egg during the process of fertilisation and the formation of a zygote.
- label the gametes to show their adaptations to their roles in fertilisation.

KEY STAGE 4 CONCEPTS

Gametes and fertilisation are considered briefly at KS4. Formation of gametes is discussed in Chapter 3 and their role in increasing variation in species is discussed in Chapter 19.

KEY STAGE 3 CONCEPTS

Reproductive anatomy

Pupils now know that sexual reproduction requires the fusion of a sperm cell and an egg cell in fertilisation. To bring the sperm and egg together, humans have a complex reproductive anatomy, each part of which is either involved in the production of sperm or eggs (ovaries and testes), or in bringing them together (penis, vagina, uterus and oviduct). The uterus, or womb, is also designed to house the developing foetus. Bear in mind that, in common parlance, semen is referred to as sperm. Stress that there are actually millions of sperm within the semen (which also contains secretions from the prostate gland and seminal vesicle). Provide pupils with diagrams of the male and female reproductive organs to label. They should understand the broad functions of each part of the reproductive system.

- Your department may have a video which explains the functions.
- You could discuss them as a class.
- Pupils could research the functions from books or CD-ROMs.

You should include:

- Male: testes, epididymis, scrotum, sperm duct (vas deferens), urethra, penis, seminal vesicle and prostate gland.
- Female: vagina, cervix, uterus (womb), oviduct (Fallopian tube) and ovary.

Eltringham and Lock (1998) suggest that pupils are often not encouraged to relate the diagrams of the reproductive organs to their own bodies. If you feel confident with your group, pupils could draw the reproductive organs into the correct position on cookery aprons.

Assessing pupils' learning

- Give pupils a quiz to test their knowledge of human reproductive anatomy.
- More able pupils could write about the functions of the reproductive system. Less able pupils could match up cut-out boxes of the functions with the labels on their diagrams.

The menstrual cycle

Your group will have a range of knowledge about the menstrual cycle. Some girls may have started their periods, some may not. Some boys will have been told about the menstrual cycle by parents, friends or siblings.

The menstrual cycle arises because of the need for eggs to be moved to the site of fertilisation, the oviduct. When a girl is born, she has hundreds of thousands of eggs in her ovaries. At some point between the ages of 9 and 18, release of these eggs begins. If you refer to the age at which the menstrual cycle should begin, stress that it varies between individuals. The release of an egg by an ovary is called ovulation, and this happens *approximately* once every 28 days. Stress that 28 days is just an average figure, and that cycles will differ both between individuals and between months.

Once the egg is released, it is wafted by tiny hairs (cilia) in the oviduct funnel into the oviduct. While this is happening, the lining of the uterus becomes thick with blood vessels. If the egg is fertilised in the oviduct, it will pass down into the uterus and embed itself in the uterus wall. If not, it will pass down to the uterus, and the uterus wall will break away and pass out of the body with the unfertilised egg. This is the period. The loss of blood from the body is called menstruation; hence, the menstrual cycle. While the new egg is ripening in the ovary, the lining of the womb thickens again, ready to receive another egg.

The easiest way to represent the menstrual cycle is that shown in Figure 10.1. Take note that by convention, day 1 of the menstrual cycle is not the day on which the egg is released from the ovary; it is the first day of the period, or menstruation.

73

Assessing pupils' learning

Ask pupils to

- arrange the pictures in the correct position around the disc shown in Figure 10.1 and to explain what is happening at each point.
- give advice to a couple who have not yet conceived, but have been trying to have children. When should they have sexual intercourse to stand the maximum chance of fertilisation?
- make a concept map, linking aspects of male and female reproductive anatomy, fertilisation and the menstrual cycle.

KEY STAGE 4 CONCEPTS

The menstrual cycle is included at KS4, but the focus is on how hormones control that system (see Chapter 12).

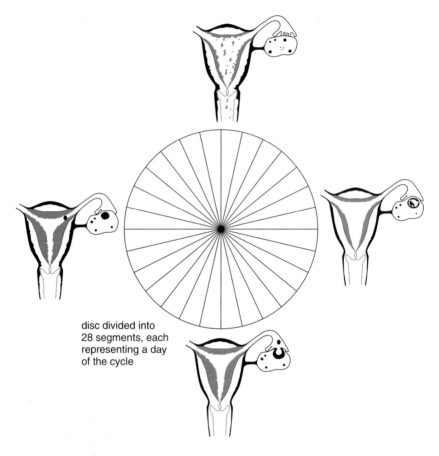

disc divided into
28 segments, each
representing a day
of the cycle

Figure 10.1 *The menstrual cycle*

KEY STAGE 3 CONCEPTS

Copulation

Your department is likely to have a video which describes copulation. If so, devise a question sheet which allows you to bring out the important points; namely:

- The penis delivers sperm into the female's vagina.
- Before the penis can be inserted it must become stiff and erect by being swollen with blood. Stress that the penis has special erectile tissue into which the blood flows.
- Release of sperm (ejaculation) is controlled by hormones. Sperm is squeezed from the testes, through the sperm duct and out through the urethra by muscular contraction.

- Having been released into the female's vagina, the millions of sperm released in one ejaculation swim through the cervix and the uterus into the oviduct. If there is an egg there, it is likely to be fertilised.

How much you discuss copulation as love-making is up to you, and will depend on your school's sex education policy. The subject may be easier to deal with if you treat it simply as a physiological process, and insist on referring to it as copulation. If you want to, or must discuss copulation as part of love-making, then deal with the purely physical aspects of copulation first, then move onto the role of sex in love. Take advice from the PSE co-ordinator about how to approach the subject effectively.

Assessing pupils' learning

- If you have used a video and question sheet, assess understanding from their answers.
- Ask pupils to write in the first person, describing the journey of a sperm cell from the testes to fertilising the egg in the oviduct.

Fertilisation and implantation

Having mentioned both of these earlier, you can discuss them in greater detail with more able pupils. When the sperm cells meet the egg, only one of them will successfully break the egg membrane and get inside. As soon as a sperm has entered, the egg lays down a thickened membrane to prevent any other sperm getting in.

Ask pupils how many cells are in humans, and how we got so many, having only started with one (the zygote). Most will realise the zygote must have divided repeatedly.

As it divides, the ball of cells continues to move down the oviduct until it reaches the uterus where it embeds itself into the uterus wall. Seven days are required between fertilisation and implantation. The embryo uses enzymes to digest its way into the uterus wall, and is supplied with food and oxygen by the blood vessels in the thickened lining.

Assessing pupils' learning

Ask pupils

- to annotate a diagram showing the different stages the developing zygote and embryo go through prior to implantation.
- why the embryo has to implant itself into the uterus wall: to gain access to food and oxygen through the blood.

KEY STAGE 4 CONCEPTS

Fertilisation is included at KS4 in relation to inheritance. This is discussed in Chapter 20. Implantation is not included at KS4.

KEY STAGE 3 CONCEPTS

Pregnancy

Pregnancy is the period between fertilisation and birth. In humans, it lasts about nine months, during which time the embryo will continue to divide and its cells begin to specialise. After two months, it becomes too big to be contained within the uterus wall, becomes known as a foetus, and floats free in a bag of fluid (amniotic fluid) in the uterus. This bag of fluid helps to protect the baby from knocks. You can demonstrate the role of the amniotic fluid using eggs. Put an egg into a bag full of water and seal it, expelling any air. Drop this from 50 cm height. To ensure the egg does not break, hard-boil it before the lesson (do not tell the pupils!). Place a normal egg into a bag without any fluid and drop it from 50 cm height. This one should break.

The foetus remains connected to the uterus wall by an umbilical cord. This supplies the foetus with food and oxygen from the mother. You need to explain how the embryo gets its food whilst in the uterus. Ask pupils what people need to survive. Most will suggest food (which you eat) and oxygen (which you breathe). If they have studied respiration at KS3, they should be able to explain why the embryo needs food and oxygen. Since the foetus cannot breathe or eat, it must absorb its food and oxygen from the mother through the blood. Confirm with pupils that they need to expel carbon dioxide (through the lungs) and waste (through faeces and urine). Again, suggest that the foetus cannot breathe out, urinate or defaecate and that waste is removed from the foetus through the blood to the mother, who will then expel it from her body.

Although the embryo can exchange food, oxygen, carbon dioxide and waste with the mother's capillaries in the uterus wall, as it continues to divide and get bigger it needs to become more efficient. To increase the surface area over which materials can be exchanged, a placenta develops. This is made up partly from the embryo's tissue, and partly from the mother's tissue. Within the placenta, the blood of the baby and mother do not mix, but pass extremely close to each other. This allows exchange of materials to occur between the two sources of blood.

Ask pupils why a mother's and baby's blood should not mix: (i) the mother's blood pressure is too high for the baby's blood vessels to withstand, and (ii) blood separation may limit the spread of some diseases.

Finally, explain how an increase in surface area increases the efficiency with which substances can be exchanged. Remind pupils about the drainpipe demonstration (see Chapter 6) and the practical showing the importance of surface area (see Chapter 7). Less able pupils' understanding of the function of the placenta is often very poor. Use clear, schematic diagrams to help.

Assessing pupils' learning

- Ask pupils to rearrange diagrams of the baby at different stages during pregnancy to represent its growth and development.
- Ask pupils to label a schematic diagram of the placenta and developing baby.
- Pupils should demonstrate written understanding of the functions of the placenta. Namely, which substances are exchanged through the placenta, and how the placenta is adapted to such exchange.
- Pupils should write up the 'egg' experiment, concluding correctly why the baby should be kept in a bag of fluid.
- Ask pupils to write a pamphlet for pregnant mothers, explaining why it is dangerous to smoke or drink alcohol during pregnancy.

Birth

Your department should have a video about birth which it considers appropriate. Ensure pupils are seated on chairs and keep a close eye on them: some may feel nauseous or even faint!

Birth is a difficult process, not least because a large baby has to escape through the cervix and vagina, both of which are narrow openings. Pupils should predict that birth will be easier if the baby is born head first, and that the cervix and vagina will have to increase in diameter. Pupils may have heard of 'contractions' before. If your class has studied muscles and movement, they should realise that contractions are carried out by muscles in the uterus wall.

There are two phases of contractions. The first phase opens the cervix; the second phase pushes the baby down. Before the baby passes out of the uterus, the amnion (the sac holding the fluid which surrounds the baby) breaks and the fluid passes out through the vagina. This is referred to as the 'waters breaking'. A woman is said to be 'in labour' once this has happened. When the baby has passed out of the body, it is still attached to the placenta with the umbilical cord. Pupils should be able to suggest that the baby needs separating from the placenta as it can now survive on its own. The umbilical cord is usually clamped before it is cut, in order to minimise blood loss and prevent infection. A short while after this, the placenta (the afterbirth) passes out of the body with more contractions.

Growth and development

Reproduction does not just involve the production of a new baby. That baby has to develop into an independent adult which has the ability to reproduce itself. The changes in the body which allow this to happen are referred to as development. Growth is the increase in body size which accompanies development. Ask pupils how the zygote grew into an embryo: by cell division. Stress that all living things grow by cell division, even after they have been born.

Line up your class in height order and pupils will see that girls are generally taller than boys. Use this result to suggest that males and females grow at different rates. Ask pupils to plot growth curves for males and females between the ages of 0 and 20. Pupils could enter the data into a spreadsheet and plot the graph on the computer. More able pupils will realise that the steeper the graph, the quicker the growth, and should realise between which ages growth is fastest for both males and females. You may need to give less able pupils the graph already plotted. Ask pupils to identify when males grow quickest, when females grow quickest, when females are, on average, larger than males, and when males are, on average, larger than females.

There are several stages of growth and development: post-natal (0–1 years), infancy (1–5 years), childhood (5–11 years), adolescence (11–18 years), adulthood (18–65 years), old age (65–death). The body undergoes changes in each of these periods. The changes which occur in adolescence are those which are most important to becoming independent adults and being able to reproduce. Depending on the maturity of your group, you could divide them into small, mixed-sex groups, and ask them to make a list of the changes which occur during each stage of development.

Having completed this, explain the difference between adolescence and puberty. Puberty is the development of secondary sexual characteristics (those changes where the sex organs begin to produce gametes, and the physical appearance of the male and female body changes). Adolescence refers to the period of life when puberty occurs, but also refers to the developing mental and emotional maturity needed for adulthood. For example, adolescents start to think for themselves, and rebel against the opinions of their parents. Pupils could devise short role plays demonstrating some of the problems of adolescence.

Assessing pupils' learning

- Ask pupils to define growth and development.
- Pupils could answer questions about growth rates at different ages.
- Give pupils drawings of sexually mature men and women. Ask them to identify secondary sexual characteristics. Alternatively, give them a list of the changes which occur at different stages of development, and ask them to sort them into appropriate age groups.

KEY STAGE 4 CONCEPTS

Growth and development are not included explicitly at KS4. Hormonal control of growth is considered briefly in Chapter 12.

References

Henderson, J. (1994) Teaching sensitive issues in science: the case of sex education. In: *Secondary Science: contemporary issues and practical approaches* Wellington, J. (Ed.), pp 240–257. Routledge, London

Eltringham, J. & Lock, R. (1998) Amazing aprons – an attractive, alternative anatomical aid. *Journal of Biological Education* 32:7–11

Nervous control

BACKGROUND

Humans, like other higher animals, have a nervous system. This controls our bodies and allows us to react to external stimuli. There are two main parts: the central nervous system (CNS: the brain and spinal cord) and the peripheral nervous system (PNS: the rest of the nervous system). The nervous system is made up of lots of nerve cells (neurones). Messages are sent along sensory neurones from receptors in the PNS to the CNS with information about what is happening in different parts of the body or outside the body. Such receptors include the eyes, ears, skin, nose and tongue. The CNS acts on this information by sending a message to an effector along a motor neurone. For example, a sense cell in the skin of the hand may detect that a plate it is holding is hot. The spinal cord receives this information and sends a message to a muscle in the arm to drop the plate. Responses like these which do not involve conscious thought are known as reflexes and the pathway of the nerve impulses involved is known as a reflex arc.

The nervous system is not included at KS2 or 3, but is included at KS4 double and single award. You should be aware that pupils will have developed their own understanding of how their brains, eyes and ears operate. They sometimes find it difficult to realise that receptors merely take in information and send it to the brain. As far as they are concerned, the eye is solely responsible for seeing and the ear is solely responsible for hearing. They are often unconvinced that the brain actually forms the picture and sound based on the information it receives.

KEY STAGE 4 CONCEPTS

Sensitivity

Sensitivity is first introduced to pupils at this stage. Begin by asking why animals need to react to external stimuli: (i) to avoid predators, (ii) to get food, (iii) to find a mate, (iv) for protection. Ask pupils to make a table of stimuli, the sense organ which reacts to each stimulus, and the sense associated with that response. They should include light (eye, sight), sound (ear, hearing), pressure (skin, touch), pain (skin, touch), temperature (skin, touch), chemicals in the air (nose, smell) and chemicals in food (tongue, taste).

Pupils can investigate which region of their tongue is sensitive to what sort of taste (sweet, sour, bitter, salty). Get pupils to dry their tongues and then, using straws, drip each of different solutions (sugar solution, vinegar, black tea and salt solution) on the side, tip, middle and back of the tongue (dip the straw into the solution, place your finger over the top, and lift the straw, with a drip of solution within it). The tip of the tongue should be sensitive to sweet and salt, the sides to sour, the middle to sweet and the back to bitter.

S **Safety Advice:** Ensure that only sterile straws are placed into the stock tasting solutions. Straws should be placed in absolute ethanol after use.

Pupils may not immediately identify the skin as a receptor. There are different types of receptor within the skin. Remind pupils about going to the dentist and having an injection. Although they can feel no pain, they can still feel a sensation resulting from touch; the dentist has stopped the pain receptors from working, but has left the pressure receptors operative. You can do three investigations of the receptors in the skin: one to establish the distribution of pressure receptors on the hand (Figure 11.1a), one to work out the distance between receptors (Figure 11.1b), and one to work out the distribution of heat receptors on the hand (Figure 11.1c).

S **Safety Advice:** Warn pupils not to stab the pins too hard into the skin.

When you conclude these practicals, ensure you introduce the idea that receptors are connected to the brain by nerve cells or neurones. These are effectively like wires, transmitting electrical impulses around the body. When two of the pins are close together, they are actually hitting the same receptor, and therefore only one message goes back to the brain.

Having established that the sense organs are connected to the CNS by nerve cells, remind pupils that we detect stimuli in order to react to them. Build up a 'stimulus–response' model of how we react using the practical in Figure 11.2. The stimulus is movement of the ruler. The receptors are the eyes. A message

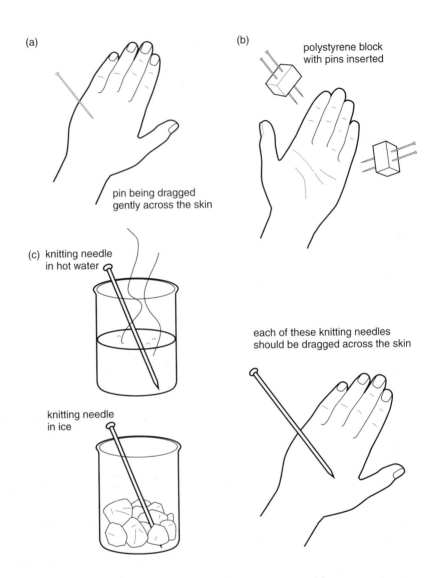

(a)

pin being dragged
gently across the skin

(b) polystyrene block
with pins inserted

(c) knitting needle
in hot water

each of these knitting needles
should be dragged across the skin

knitting needle
in ice

Figure 11.1 *Conduct these experiments on a subject who is blindfolded or looking away. a) Apparatus for detecting the distribution of pressure receptors in the skin. Very gently stroke the skin with the pin. Record the areas of the hand which recognise the pin's pressure. b) Apparatus for establishing the distance between pain receptors in the skin. Press the pins against the skin 10 times. Then change the distance between the pins and repeat. In each 10 stabs, stab five times with just one pin and five times with both pins. For each distance between the pins, record the percentage of stabs which were identified accurately as one or two pins. When the pins are 1 cm apart, it is easier to recognise both pins. When the pins are 1 mm apart, it is more difficult to recognise both pins. c) Apparatus for detecting the distribution of heat receptors in the skin. Record those regions of the hand which accurately identify whether the needle is hot or cold*

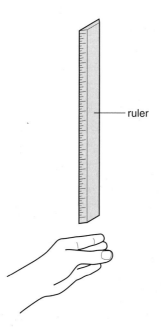

ruler

Figure 11.2 *Apparatus for investigating the stimulus-response model of behaviour. Hold the ruler just above someone's hand. Without telling them, drop the ruler and measure how far it drops before being caught*

passes from the receptors along the sensory neurone to the brain. The brain then sends a message along the motor neurone to your muscles (effectors), and you catch the ruler (response).

- Ask pupils to construct a generalised flow chart outlining this model of response.
- Reinforce this stimulus–receptor model using role play. Use individual pupils to represent the receptor, central nervous system and effector. Lines of pupils can connect them up (they are the neurone) and pass messages between them. Use a balloon with a message taped to it to represent the message travelling through the neurone. Pupils should hold the balloon between their knees and pass it along the line.

Extend the activity in Figure 11.2. Make the 'catcher' close their eyes, but position the fingers to just touch the ruler. You can calculate the reaction time according to the formula: $t(\text{seconds}) = \sqrt{(\text{distance}/10)}$. Pupils could use a spreadsheet to calculate reaction times. The reaction time with closed eyes will be longer than with open eyes. This is because the impulse from the touch sensors in your skin has further to travel to the brain, than the message from your eyes.

- Pupils should write up all the investigations, explaining their results as summarised above.
- Pupils should demonstrate that they understand the location of receptors. Try drawing cartoons of animals, exaggerating those sense organs which they think are most important to each species. For example, a cartoon rabbit may have incredibly large ears, and eyes poking out on stalks.
- Give pupils different situations where they can identify a stimulus, and for which they can define a response. Ask them to draw a flow chart showing the pathway of nerve impulses in each case.

The structure of the nervous system

Distinguish between the central nervous system (CNS) and peripheral nervous system (PNS). Give pupils a diagram showing the brain, spinal cord and other nerves, and ask them to indicate that the spinal cord and brain are part of the CNS and the other nerves part of the PNS. As extension work, higher level pupils could research the functions of different regions of the brain from textbooks or CD-ROMs.

Use questions to help pupils think of the adaptations of the 'wires' that comprise the nervous system. Think about how wires are arranged to supply electricity to each room of a house, and to each part of each room. Bring out the following adaptations: insulation (to make the message travel quickly), branched in the brain (to receive information from lots of different neurones), branched in the effector (to make every part of a muscle contract, for example), and long (to reach all the way around the body).

Pupils may need to know the structure of a sensory neurone and of a motor neurone. You can find these in most textbooks. Ask them to draw them, and to indicate the adaptations of each. Many pupils do not realise that neurones are single cells. The idea of an animal cell being spherical sticks in their minds. The cell body of the neurone is indeed spherical – it contains most of the contents of the cell. However, the long processes arising from the cell body are also part of the cell and contain cytoplasm. The names of these processes depend on the direction of the nerve impulse: the axon carries impulses away from the cell body; the dendron carries impulses towards the cell body.

To connect up to other neurones, there must be a junction across which nervous impulses can pass. This is called a synapse. More able students should be aware of its structure and function; the detail depending on your syllabus. The synapse is a microscopic gap between the ends of the neurones. When a nervous impulse arrives at the end of the 'so-called' pre-synaptic neurone, it releases a chemical (a neurotransmitter) which crosses the gap and causes a nerve impulse to move off down the post-synaptic neurone.

You could modify the earlier role play to demonstrate this. Form your class into a line, but leave a 3 m gap halfway down each line. This gap is the synapse. Pupils must again pass the message taped to a balloon along the line. When it reaches the synapse, show pupils that throwing the balloon across will not work: it will not reach the other side of the gap. This mirrors the synapse: an electrical message will not cross the gap. Instead, pupils should remove the message from the balloon, tape it to a tennis ball, and throw it across the gap (the tennis ball represents the chemical neurotransmitter). It is impossible to pass a tennis ball from person to person between the knees. Therefore, the message must be taped to a balloon again (i.e. translated back into an electrical impulse).

Assessing pupils' learning

- Ask pupils to identify the central and peripheral nervous systems on diagrams of each.
- Ask pupils to write about the adaptations of neurones.
- More able pupils could produce a diagram of the synapse with annotations to explain what happens when a message arrives at the pre-synaptic neurone.

85

Reflexes

Reflexes are responses which are fast, protective and automatic; they do not require conscious thought. There are some simple experiments to demonstrate reflex actions:

- Gently blow air into a partner's eye; the eye-lid will close.
- Using a lamp and mirror, shine light into a partner's eye; the pupil will close automatically.
- Drop some concentrated lemon juice onto a partner's tongue; their face will 'screw-up'.
- Sit on a bench with crossed legs. If a partner taps just below the knee cap, the lower leg should spring up.

More able pupils should be aware that there are two types of reflex: the spinal reflex and the cranial reflex. Most syllabuses only require knowledge of the spinal reflex. A spinal reflex response which features most often in examination questions is that shown by the hand when it touches a hot plate. You can find a diagram of the reflex arc associated with this response in most textbooks. Talk through the path of the nervous impulses. Also tell pupils that a message will be passed to the brain so that it is aware of the response that has happened.

Most able pupils may ask how we know to move our hands away from the cooker, a split second before our hand actually hurts. This is because the reflex arc takes less time than it takes for a message to be passed back to the brain.

As extension work, more able pupils could learn about Pavlov. He investigated conditioned reflexes. He noted that dogs always began to salivate when presented with food. He sounded a bell just before giving them their food, and after many repetitions, he found that the sound of the bell alone promoted salivation. Pupils could investigate whether human reflexes can be conditioned. Let pupils blow gently into each other's eyes using a straw. The recipient blinks in response. Each time they do so, a bell should be sounded. After at least 30 repetitions, they should hold the straw up to the person's eye and ring the bell without puffing air into the eye. The pupils' eye will still blink.

Assessing pupils' learning

- Ask pupils to write up the reflex experiments, describing the survival value of each response.
- Pupils should construct a table, indicating the stimulus, receptor, response and the effector which caused that response.
- Pupils should be able to label a diagram of the reflex arc and indicate the direction of movement of the impulse.

The eye

To introduce the eye, begin with an analogy of a pin-hole camera. Pupils can make these using tin cans, black paper and greaseproof paper screens (Figure 11.3). If they make a small pinhole in the black paper and point the black paper at a lamp, they should see the image of the lamp on the screen. It is easier to see in a dark room.

Pupils should conclude that a camera uses light to form an image of an object. Tell pupils that eyes also use light to form an image of an object. Many less able pupils will be resistant to this idea. Because the eye is doing the seeing, it is unintuitive that something (light) comes into the eye to allow sight to occur. Many younger pupils think of 'something' going from the eye to the object. Try to overcome this misconception as follows:

- Tell them to shine a torch into a mirror and look into the mirror at the torch's reflection. Ask them in which direction light travels when they see the lamp in the mirror.
- Challenge them to explain why a camera cannot form an image of an object in the dark: because there is no light available. The camera has not changed, but

You will need

aluminium foil (about 5 cm × 5 cm)
large nail
pin
empty tin with a base about 120 mm square, no lid is required
hammer
greaseproof paper (tracing paper)
black sugar paper
Sellotape

How to make the camera

1 Draw a line with a pencil across the closed end of the tin.
2 Draw another line across the tin so that it crosses your first line at right angles.
3 Use the hammer and nail to punch a hole in the tin at the place the two lines cross over.
4 Tape aluminium foil over the hole.
5 Use the pin to make a very small hole in the aluminium foil. This should be in the middle of the hole in the tin.
6 Tape the greaseproof paper over the open end of the tin.
7 Point the end of the tin with the hole in it at a window at least 5 metres away. Look at the greaseproof paper end. You should see an image of the window.

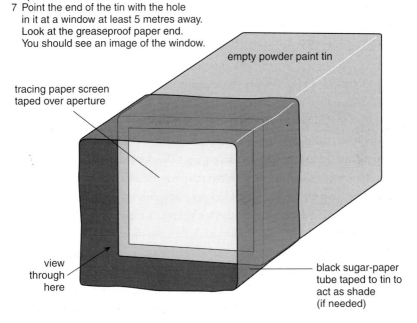

front

strip of aluminium foil to act as a shutter

make pinhole here

empty powder paint tin

tracing paper screen taped over aperture

view through here

black sugar-paper tube taped to tin to act as shade (if needed)

Figure 11.3 *A pinhole camera*

the light in the environment has, suggesting light passes from the object to the camera (and likewise to the eye).

- Some pupils, particularly in towns, will suggest that we can see in the dark, and therefore that the eye must produce something which allows us to see. This

87

misconception usually originates because few children experience total darkness: remind them of this and that street lamps or the stars actually provide light which the eye uses to see at night.

It is perhaps more difficult for pupils to appreciate why light should pass from, for example, a book into the eye. The light source is obvious when using a lamp or a torch. Remind them that the Sun provides us with light, and that light reflects off the book into our eyes.

Provide pupils with a diagram of the eye to label from a textbook. Dissect a pig's eye as a demonstration, or allow pupils to dissect eyes in small groups. Try to slice the eye so you cut through the middle of the iris. Ensure pupils can relate the diagram to the dissection.

S **Safety Advice:** If you carry out the dissection as a class practical, check current DfEE regulations and with your head of department. Avoid using sheep's or cows' eyes. If you cannot do the dissection at all, an interactive cow's eye dissection is available on the internet at http://www.exploratorium.edu/learning_studio/cow_eye/

Either during or after the dissection, relate the eye to the pinhole camera, and trace the pathway of light through the eye. Your department may have a model eye which you can use to help with this. You can compare the retina to the greaseproof paper screen (where the image is formed) and the pupil to the pinhole (where the light enters).

Establish the differences between the camera and eye – ask pupils what the eye's lens is for; many will know from physics that it focuses light. Demonstrate how a lens can focus an image by setting up a slide projector (Figure 11.4). Pupils will need to experiment with the position of the lens to get the image focussed on the screen. Note the image is upside down and that the image on the retina is also upside down. Conclude that the lens focuses the light and light needs to be focussed to create a sharp image on the retina. Mention that the cornea also focuses light in the eye (in fact, it does most of the focussing).

Cameras and eyes do differ. In a camera, light is focussed onto the film, and a permanent image recorded. The image projected onto the retina is only interpreted into a picture by the brain. Remind pupils that for your brain to produce an image, information from the retina has to be passed back to the brain along the optic nerve (you can show this to pupils at the back of the eye). It is like a *digital* camera, where information must be passed to a computer to be interpreted into a picture. To convince pupils that the brain is required to interpret the information from the eye remind them of the following:

• If the brain did not interpret the image then we would see upside down!
• If optical illusions (Figure 11.5) can be interpreted in more than one way,

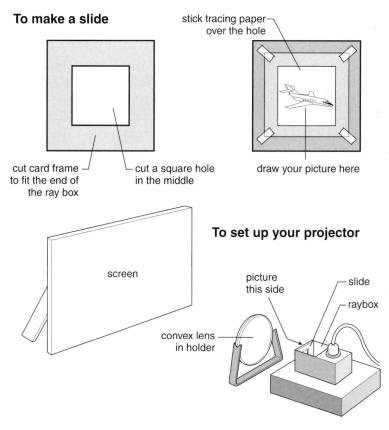

To make a slide

cut card frame
to fit the end of
the ray box

cut a square hole
in the middle

stick tracing paper
over the hole

draw your picture here

To set up your projector

screen

picture
this side

slide

raybox

convex lens
in holder

Figure 11.4 *A slide projector*

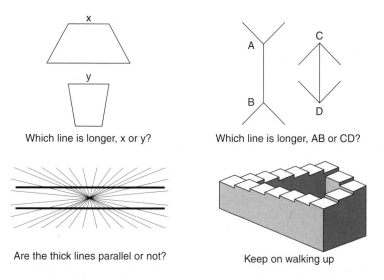

x

y

Which line is longer, x or y?

A

B

C

D

Which line is longer, AB or CD?

Are the thick lines parallel or not?

Keep on walking up

Figure 11.5 *Optical illusions*

something (which is the brain) must be doing the interpreting! There are a wide range of optical and sensory illusions included on the Illusionworks website (http://www.illusionworks.com/).

- Hallucinogenic drugs distort what we see. They do so because our picture of the world depends on the brain's interpretation of the nervous impulses it receives. Drugs affect this interpretation.

More able pupils will realise that the retina appears black on dissection because it absorbs light. Certain syllabuses require more able pupils to know some detail about the retina. The retina has light sensitive cells, rods and cones, which send messages back along the optic nerve to the brain where it interprets the messages into an image. Rods are not sensitive to colour, but work in dim light. Cones detect colour.

To reinforce the ideas of light being focussed onto the retina, the brain creating an image by interpreting messages from the retina, and the optic nerve leaving the back of the eye, try the following:

- Tell pupils that the point at which the optic nerve leaves the retina is called the blind spot. Because the optic nerve leaves at this point, there are no light sensitive cells on this part of the retina.
- Ask pupils to draw a dot on a piece of paper, and about 10 cm to the right, draw a cross. They should hold the paper at arm's length, cover their right eye, and look at the cross with their left eye. If they bring the paper closer to their eye, and keep focussed on the cross, the dot will disappear.
- This is because the image of the dot is focussed onto the blind spot, and therefore does not stimulate any light sensitive cells in the retina. As a result, no information about the dot is sent to the brain, and the dot does not form part of the image interpreted by the brain.

Your discussion of how light is focussed by the eye will depend on your syllabus, pupils' prior knowledge from physics, and on their ability. With less able pupils, avoid drawing ray diagrams. Simply tell them that the cornea and lens bend the light to focus it. Higher level pupils may need to draw ray diagrams.

More able pupils may need to understand how eyes can adapt to high and low light intensities. Ask pupils to work in pairs. One member of the pair should close one eye for 20 seconds. Upon opening, the pupil will have expanded. When it is opened, the iris reduces the size of the pupil immediately (a response which a partner looking at the eye should be able to see) to protect the retina from excessive light. Provide pupils with diagrams showing the location of radial (point outwards from the pupil like the spokes of a wheel) and circular muscles in the iris (concentric rings of muscle around the pupil), and ask them which must contract to dilate the pupil (dim light; radial muscles) or constrict the pupil (bright light; circular muscle). Remember if one type of muscle is contracted, the other is relaxed. If they find this difficult to understand, use the model described in Figure 11.6 (Lock 1992). Point out that when the circular muscles are contracted (the ring is made smaller), the radial are relaxed (get longer), and vice versa.

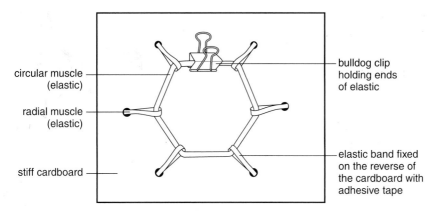

Figure 11.6 *A model of the circular and radial muscles in the iris of the eye*

Some syllabuses require higher level pupils to understand the detailed effect of the lens on the fine focussing of light. The lens can change shape according to whether the ciliary muscles are contracted or relaxed. When they are contracted, the suspensory ligaments slacken and the lens becomes fat. This allows it to focus light more effectively from close objects. When the ciliary muscles are relaxed, the suspensory ligaments tighten and the lens becomes thinner. This is more effective at focussing light from more distant objects. You can modify the projector experiment in Figure 11.4 to include different shaped lenses. The most able pupils could draw ray diagrams to show the effect of different thicknesses of lens, both in the projector and in the eye. The process by which the lens changes shape is called accommodation.

As extension work, more able pupils could consider why most animals have two (rather than one) eyes. They should design investigations to test appropriate hypotheses; for example, (i) to judge distance more accurately, or (ii) to provide a wider field of view.

Assessing pupils' learning

- Pupils should label a cross-section of the eye.
- Ask pupils to describe the pathway of light through the eye using a flow chart.
- Pupils should make a table describing the function of different parts of the eye.
- Pupils should write up the blind spot experiment, explaining why the spot disappears.
- Pupils should explain how the eye adapts to different light intensities.
- Ask pupils to research how short- and long-sightedness are caused (use older biology textbooks).

References

Lock, R. (1992) Antagonistic muscle action – modelling the control of the iris. *Journal of Biological Education* **26**:164–165

chapter 12

Hormonal control

BACKGROUND

As well as the nervous system, animals also have an endocrine system (hormonal system), made up of endocrine glands, which helps to control the body. The pituitary gland in the brain controls the release of hormones from other endocrine glands. It also produces growth hormone and anti-diuretic hormone (ADH), which controls water levels in the blood. The pancreas controls blood sugar levels, the testes and ovaries control sexual development and the production of gametes, and the adrenal glands control the body's responses to emergencies. The hormones released from these glands are released into the blood and act only on specific cells. Because hormones are carried through the blood, the body's responses to stimuli are not as quick as those produced by the nervous system. Hormones can be used artificially in fertility treatment and in the treatment of diabetes.

The hormonal system is not included at KS2 or 3. It is included at KS4 double and single award. If you have already covered hormones in plants, pupils will have some understanding upon which to build. Pupils may have heard of the sex hormones before. Diabetics, or relatives and friends of diabetics, will be familiar with diabetes and the roles of insulin and glucagon. Be aware of diabetics in the class and exercise sensitivity.

KEY STAGE 4 CONCEPTS

The endocrine system

Begin by suggesting that while the nervous system allows the body to perform immediate responses to external stimuli, in general, the hormonal system controls longer-term changes in the body. There are two types of gland.

Brainstorm the glands with which pupils are familiar and divide them up as you write them on the board. Exocrine glands (e.g. the tear gland and some parts of the pancreas) secrete their products directly to their place of action (tears and digestive enzymes, respectively). Endocrine, or ductless, glands secrete their products into the blood plasma which carries them to their place of action. Hormones are produced by endocrine glands. Pupils should be able to identify and locate the major endocrine glands. Most textbooks have good diagrams of the endocrine system.

Nervous impulses pass around the body through nerve cells. Hormones do not have their own routes around the body; they are secreted into the blood. Specific hormones, although passing all around the body to every cell, only act on particular target cells. Such cells have receptors to these hormones on their surface. Therefore, the action of hormones is just as specific as the action of nerves.

Assessing pupils' learning

- Pupils should produce a table to compare and contrast the nervous and endocrine systems. Focus on the nature of the message (chemical/electrical), the source (gland/receptor or CNS), how it is carried (bloodstream/nerve cell), speed (slow/fast) and duration of effect (long-lasting/brief).
- Ask pupils to distinguish between a list of endocrine and exocrine glands.
- Pupils could write in the first person, from the point of view of a hormone molecule, explaining how it reaches and recognises its target cells.

Control of blood sugar levels

Begin by asking why the body needs to control the glucose levels in the blood. Many pupils will remember that glucose is used to provide us with energy, and that the blood transports glucose to the cells throughout the body. If the blood does not carry enough glucose, the body cells will not receive enough for respiration. If more able pupils have studied osmosis, ask them to predict what would happen to body cells if the concentration of glucose in the blood was too high. Water would be drawn out of the body cells into the blood and stop them working properly (see Chapter 4).

More able pupils must know that blood glucose levels are controlled by negative feedback (see Chapter 13). Build up the following as a flow chart on the board:

- If blood sugar levels get too high, insulin is released from regions of the pancreas called Islets of Langerhans. When insulin arrives at target cells in the liver, the cells convert glucose in the blood to glycogen which is stored.

- When blood glucose concentration falls, the pancreas stops producing insulin.
- If blood glucose concentration gets too low, the Islets of Langerhans release a hormone called glucagon. When this arrives at the liver cells, it causes them to break down glycogen and to release glucose into the blood.

Assessing pupils' learning

- Ask pupils to produce a flow chart showing how blood glucose level is controlled.
- Give pupils a graph showing how blood glucose level varies during a day (Figure 12.1). Pupils should explain peaks and troughs in the graph as the result of meals and the action of insulin and glucagon.

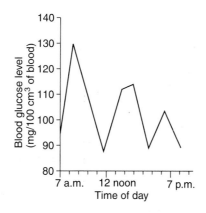

Figure 12.1 *Fluctuations in blood glucose levels over one day*

Diabetes

If the pancreas does not produce enough insulin to regulate blood glucose level, a person will suffer from *diabetes mellitus*: sugar diabetes. Your department may have information leaflets about diabetes. Ask pupils to research symptoms and treatment. Symptoms include (i) constant thirst, (ii) general weakness, (iii) loss of weight, (iv) glucose present in the urine, (iv) possible coma followed by death.

Make two beakers of cold tea (to simulate urine samples) and dissolve sugar in one. Ask pupils to test the solutions for sugar (see Chapter 6), and identify which sample comes from a diabetic. You can tell pupils that doctors used to test urine for diabetes by tasting it. You can cause uproar in the classroom by dipping your finger into the 'urine' and tasting it!

More able pupils could explain why the body develops a constant thirst: if there is a high concentration of glucose in the blood, water will move out of body cells into the blood by osmosis (see Chapter 4), and as a result, will be lost from the body in the urine (see Chapter 13).

Assessing pupils' learning

- Pupils should present their research findings about diabetes.
- Pupils should write up the urine practical, explaining which specimen comes from the diabetic and why.
- Many syllabuses require pupils to be aware of the treatments for diabetes. Ask pupils to think of treatments themselves: (i) control of diet – foods containing high levels of sugar and other carbohydrates must be strictly controlled; (ii) injection of insulin – it is difficult to know how much insulin to inject; if too much is injected, diabetics would need to consume extra sugar quickly.

Coping with emergencies

Remind pupils that adrenaline is released from the adrenal glands and that this helps us cope with emergencies. Most pupils will be familiar with what happens to the body when scared. Ask pupils to think of the survival value of these effects.

- How do you increase supplies of glucose and oxygen to the cells to release more energy for fighting or fleeing? The heart beat increases, breathing rate increases, glucose is released from glycogen in the liver into the blood.
- Why does the hair of most mammals stand on end when they are frightened, and therefore why do we get goose pimples? If the hair stands on end, it makes us look bigger.
- Why do the pupils of the eye dilate? To take in more light to see what is happening.

Assessing pupils' learning

Pupils could produce a leaflet or poster demonstrating the effects of adrenaline on the body, and the survival value of each of its effects.

Sexual development

Chapter 10 gives specific advice about discussing reproductive biology in the classroom. Because reproduction itself is not included at KS4, it is worth beginning by recapping reproductive anatomy from KS3, and reminding pupils about how eggs and sperm arrive in the oviduct for fertilisation. Differences in reproductive anatomy are referred to as each sex's primary sexual characteristics. At birth, there are few other differences between males and females.

For this chapter, focus on secondary sexual characteristics – those features which develop during puberty and are controlled by sex hormones produced by the ovaries and testes. These hormones are produced during puberty and also control production of the gametes: sperm and eggs.

- Ask pupils, in mixed-sex groups, to identify similarities and differences between adult men and women, hence defining secondary sexual characteristics in men and women.
- Provide pupils with a list of changes which happen between childhood and adulthood and ask them to identify whether they happen in males or females.

Tell pupils that the male hormone controlling development is testosterone, and the female hormones are oestrogen and progesterone. Stress that the age at which puberty happens varies between individuals. Textbooks which describe puberty happening in girls between 11–13 and in boys between 14–15 are merely expressing averages. Pupils should identify 'periods' as a significant difference between men and women. Remind them of the menstrual cycle from KS3. Girls have a wider knowledge of the menstrual cycle than boys. However, some girls may remain ignorant at KS4 about exactly what happens at each stage of the cycle.

Some syllabuses require higher level pupils to know in detail about hormonal control of the menstrual cycle. Most textbooks present a graph of (i) the thickness of the uterus lining, (ii) the level of oestrogen in the blood, and (iii) the level of progesterone in the blood, all plotted against time (days 1–28). From this, ask pupils to work out the effect of progesterone (maintains uterus lining during pregnancy) and oestrogen (causes uterus lining to thicken, and an egg to be released from the ovary: ovulation). Ovulation leaves a vacuole in the ovary called a corpus luteum (yellow body) which secretes progesterone. This prevents ovulation happening again. If the egg is not fertilised, the corpus luteum breaks down and progesterone levels reduce, promoting menstruation.

Assessing pupils' learning

- Ask pupils to produce a flow chart of hormonal changes and their effects during the menstrual cycle.
- Many syllabuses mention the medical uses of hormones. Explain that infertility results from a failure to produce gametes. Ask pupils to predict how males and females could be stimulated to produce more gametes (injection of oestrogen for women, and testosterone for men), and how

female contraceptive pills work (they contain progesterone which inhibits release of an egg from the ovary).

- More able pupils could research such artificial uses of sex hormones in more detail.

Growth hormone

Remind pupils they grow most during adolescence. They could plot growth curves for males and females to reinforce this (see Chapter 10). The pituitary gland makes growth hormone. This normally only has an effect during puberty. However, excessive growth (giantism) can occur if the pituitary gland produces too much hormone, or if adult cells continue to respond to the hormone. If the pituitary gland makes inadequate amounts of hormone, growth may be inadequate and dwarfism may result.

Assessing pupils' learning

- Ask pupils to predict how giantism and dwarfism may be treated: by injecting growth hormone, or surgically removing part of the pituitary gland. Ask pupils to research the role of growth hormone and the symptoms and causes of dwarfism and giantism.
- Pupils could report the findings of their research, perhaps as part of a broader project on the medical uses of hormones.
- Pupils could produce flow charts to show how giantism and dwarfism may be caused and treated.

chapter 13

Homeostasis

BACKGROUND

Humans depend on the maintenance of a constant internal environment (homeostasis). That is, humans can only survive if variables such as body temperature, and the amount of water, urea and carbon dioxide in the blood are kept relatively constant. The main reason for this is that enzymes work only at specific temperatures and pHs which can be affected by these variables. The brain controls those organs which are specialised to maintain a constant internal environment. These include the kidneys (control of water and urea levels in the blood), the skin (control of temperature) and the lungs (control of carbon dioxide levels). Control is usually by negative feedback, a mechanism which keeps variables within narrow defined limits.

Homeostasis is not included explicitly at KS2 or 3. It is included at KS4 double and single award. The National Curriculum focuses on specific aspects of homeostasis, including how waste products are removed from the body by the kidneys, how carbon dioxide is removed from the body via the lungs, how water levels are controlled by the kidneys and how humans maintain a constant body temperature. Some of these topics are also discussed in other chapters. This chapter focusses on the control of these processes.

KEY STAGE 4 CONCEPTS

Homeostasis

Pupils would benefit from breaking down the word homeostasis (*homo* = same, *stasis* = stay); it means 'staying the same'. Ask pupils what factors could change within us: oxygen, glucose, carbon dioxide, water and heat. To be kept constant, any increase in any of these factors must be accompanied by a decrease. That is:

heat gain = heat loss, water gain = water loss, carbon dioxide gain = carbon dioxide loss.

Higher level pupils must understand that the body controls these factors under hormonal or nervous control by negative feedback. You will meet this mechanism again for each of the factors described below. However, it is worth introducing the mechanism in general terms at this stage. Use the analogy of a heating system in a house (Figure 13.1). Explain that the thermometer is the receptor, and the heater is the effector. The process is called negative feedback because the change in temperature *feeds back* to the heater, causing it to act to *negate* that change.

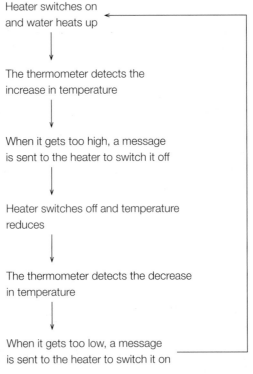

Figure 13.1 *A flow chart describing how water temperature is kept within defined limits by a negative feedback mechanism*

Assessing pupils' learning

- Pupils should explain why it is important to maintain a constant internal environment and should recall the possible factors which may vary within the body.
- Ask higher level pupils to produce a flow chart, describing how the negative feedback system works in a house heating system.

Control of blood temperature

Most pupils will say that they get warm (produce heat) whilst doing exercise.
Your department is likely to have some clinical thermometers. Ask pupils to take
their body temperature at rest, then again after five minutes vigorous exercise.
Their body temperature remains relatively constant.

S **Safety Advice:** Ensure the thermometers are disinfected, and warn pupils to be
careful whilst taking exercise. Place thermometers into absolute ethanol
immediately after use.

Pupils should conclude that their heat gain was equalled by their heat loss. Tell
pupils that their body temperature is about 37.2°C. If their temperature goes
above 40°C or below 35°C, they will die of hyperthermia and hypothermia,
respectively. You must now focus on methods by which heat is gained and lost.
Ask pupils what happens to their bodies during exercise:

1 Sweating
  - Tell pupils that glands in the skin produce sweat.
  - Show that sweat is produced using blue cobalt chloride paper pressed
    against the skin. In the presence of water it turns pink.
  - Show that sweat can remove heat from the body using the apparatus in
    Figure 13.2. Because the water in sweat has a high latent heat of
    vaporisation, it removes a lot of energy when it evaporates.

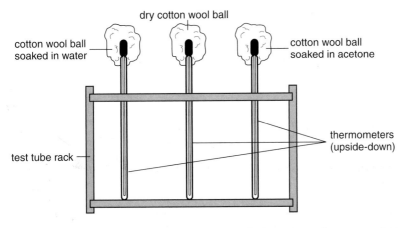

Figure 13.2 *Apparatus to show how evaporation of water removes heat energy from
the skin. Include acetone to see a particularly pronounced drop in temperature*

2 Panting
  - Pupils should breathe on blue cobalt chloride paper. It will turn pink
    indicating the presence of water (which removes heat).

- Pupils should breathe on a thermometer to demonstrate heat loss occurring through the breath.

3 Vasodilation
- Many pupils will say they get red in the face after exercise. They become red because there is an increased flow of blood to the surface of the skin. When the blood vessels near the surface of the skin dilate (vasodilation), more blood flows through them, and more heat is lost by radiation.

Ask pupils what happens when they are cold:

1 Shivering
Remind them that shivering involves your muscles doing exercise. Because of this, more respiration is happening and more heat is being produced.

2 Vasoconstriction
When cold, the blood vessels near the surface of the skin constrict (vasoconstriction), reducing blood flow, and therefore heat loss by radiation. The skin appears pale.

3 Sweating
Sweating stops, preventing heat loss.

4 Hairs stand on end
The hairs trap a layer of insulating air next to the skin, hence reducing heat loss.

5 Insulation
Many pupils will suggest that fat laid down in the skin prevents heat loss. Some pupils will know that whales have a lot of blubber (fat) to keep them warm in the cold oceans.

Some syllabuses require pupils to recognise the structure of the skin, and be able to identify sweat glands, hair, arterioles and capillaries. If your syllabus does ask for this, introduce it at this point. Diagrams are found in most textbooks.

Finally, for more able pupils you should discuss how the brain controls body temperature by negative feedback. Relate the following to the water heating system used earlier as an example. The hypothalamus (a region of the brain) is the receptor. It measures the temperature of the blood flowing through it, and it receives information from temperature receptors in the skin. The effectors are those mechanisms mentioned above (e.g. sweat glands, hairs, blood vessels).

Assessing pupils' learning
- Pupils should write up the body temperature experiment, including a prediction, concluding that body temperature is kept constant, and that heat gained must be equalled by heat lost.

- Pupils should show written understanding of the methods by which humans regulate heat loss, and of the demonstration practicals used to investigate these.
- Pupils could label and annotate a diagram of the skin, showing how each part of the skin is important for temperature control.
- Less able pupils could complete a table, showing what happens to the hairs, sweat glands and blood vessels when the body gets too hot or too cold.
- Ask pupils to design a species which is good at controlling its body heat. Encourage them to think widely about how other species control their body temperature. For instance, reptiles bask in the Sun, elephants lose heat through blood vessels in their ears, birds fluff up their feathers, and whales have large layers of fat.
- Pupils should produce a flow chart, showing how temperature is controlled by negative feedback. This could be done as a cloze exercise, sticking together components of the chart.

Control of blood carbon dioxide levels

In Chapter 5, we discussed how breathing rate changes to provide more oxygen for respiration and to repay an oxygen debt. Although this is the effect of breathing rate, its control is mediated through the brain measuring carbon dioxide levels in the blood. Remember, oxygen is used up at the same rate at which carbon dioxide is produced in respiration. Control of carbon dioxide levels is a negative feedback mechanism.

Receptors in the aorta (near the heart) and carotid body (in the neck) send messages to the medulla oblongata (a region of the brain) which itself measures the carbon dioxide levels in the blood flowing through it. The medulla then sends messages to the intercostal muscles and diaphragm to either increase (if blood carbon dioxide levels are too high) or decrease (if blood carbon dioxide levels are too low) breathing rate. The mechanism of breathing is outlined in Chapter 7.

Assessing pupils' learning

- Pupils should describe control of blood carbon dioxide levels by negative feedback. This could be done as a cloze exercise for less able pupils, sticking together phrases to form a flow chart.
- Ask pupils to explain what happens in the brain and lungs when blood carbon dioxide levels increase and decrease.

Control of water levels and excretion

Begin by remembering that water gain must equal water loss. Brainstorm methods by which water is gained (eating, drinking, a by-product of respiration) and lost (breath, urine, faeces, sweat). The body controls its water level through the kidneys.

If your department has a model human body, look at the structure of the excretory and water control system: two kidneys, two ureters, bladder and urethra. Stress that the ureter connects the kidneys and the bladder, and the urethra connects the bladder to the outside. Pupils frequently confuse these names. Pupils should label a diagram of the system.

Foundation level pupils are not likely to require detailed knowledge of kidney structure. They will need to know that the kidney acts as a filter which allows water and urea out into the urine, whereas useful molecules, such as amino acids or proteins, are retained in the blood. Provide pupils with data about the concentrations of different molecules in the blood and in the urine (Table 13.1). Ask them to explain the function of the kidney from examining these data: to expel water and urea from the body.

Table 13.1 Concentrations of substances in the blood and in the urine

	Blood plasma (g/100 ml blood)	Urine (g/100 ml blood)
Water	92	95
Protein	7.5	0
Urea	0.03	2
Ammonia	trace	0.05
Salts	0.4	1.18
Glucose	0.1	0

You could make a model of the excretory system (Figure 13.3; Newsome & Lock 1997). To operate the model, pour weak tea (straw-coloured plasma) containing red beads (red blood cells) into the funnel (the kidney). Tell pupils that the straw colour is caused by urea. The red beads will not pass through the funnel, demonstrating that red blood cells are filtered out in the kidney, and the urea passes through the ureter (the neck of the funnel) into the bladder (the balloon) which fills up. The clip at the base represents the sphincter which is only opened when someone urinates.

Higher level pupils are likely to need to know the detailed structure of the kidney. You could begin by dissecting the kidney. This is a difficult dissection, and you must practise beforehand. Examine the outside of the kidney first and

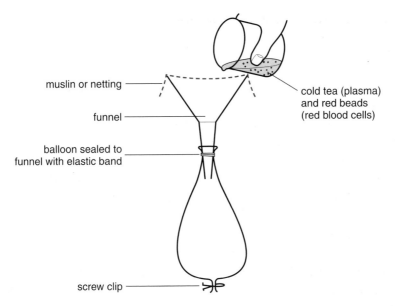

Figure 13.3 *Model of the excretory system. The funnel and netting represent the kidney and the balloon represents the bladder*

Labels on figure:
- muslin or netting
- funnel
- balloon sealed to funnel with elastic band
- screw clip
- cold tea (plasma) and red beads (red blood cells)

identify the renal artery, the renal vein and the ureter coming out from the side. Then slice the kidney longitudinally down the middle. You will probably be able to recognise regions of the kidney including some or all of the cortex, the medulla, the renal pelvis and the renal pyramids.

S **Safety Advice:** If you want the class to carry out this dissection, check the current DfEE regulations and consult your head of department. Do not use kidneys from a sheep or a cow.

To explain how the kidney filters the blood, more able pupils should understand the structure of the nephron. There are over one million nephrons in each kidney, and you can find diagrams of them, and how they relate to the broad structure of the kidney, in most textbooks. Talk pupils through the passage of molecules through the nephron. The main points include:

- Blood is forced through the glomerulus at high pressure and small molecules are forced out into the Bowman's capsule. Red blood cells and large molecules, such as proteins, are retained in the blood because they are too big to pass through the capillary walls. Small molecules and ions, including urea and glucose, pass into the filtrate.
- The glomerular filtrate passes down the convoluted tubule and loop of Henle. Numerous capillaries surround the tubules, and important molecules and

ions, including glucose and water, are reabsorbed across the tubule wall. This is known as selective reabsorption.

- Urea and water are left in the glomerular filtrate to pass into the ureter through the collecting duct of the nephron. Urea is a waste product formed when excess amino acids are converted by the liver into glucose by a process of deamination.

Osmoreceptor cells in the hypothalamus in the brain detect water levels in the blood. If the body needs to retain water, a hormone called ADH (anti-diuretic hormone) is released into the blood where it causes more reabsorption of water in the nephron. If the body has too much water, no ADH is released and water is lost to the urine.

Many pupils find this subject difficult. Johnson (1999) describes a role play which appears to have been successful in helping pupils understand kidney function. You could also ask pupils to make a class model of a nephron. Give small groups parts of the nephron to research and write up. A sieve (in the Bowman's capsule), long netting (for the collecting duct) and rubber tubing may be useful.

Assessing pupils' learning

- Pupils should be able to annotate a diagram of the nephron, explaining what happens at each part. You may need to give pupils a cloze-style exercise to complete this successfully.
- Ask pupils to produce a flow chart showing how water levels are controlled by negative feedback.
- Ask pupils to imagine themselves to be particular molecules, either water, glucose or urea, and to describe their passage through the blood and kidneys.
- Provide pupils with samples of 'urine' (Figure 13.4). Ask them to record the colour of each of these samples, and to test each sample for protein and glucose (see Chapter 6). From the colour, they should explain whether the urine samples come from a dehydrated person, or someone

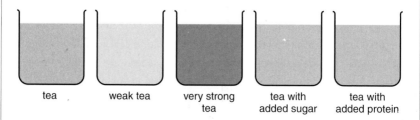

| tea | weak tea | very strong tea | tea with added sugar | tea with added protein |

Figure 13.4 *'Urine' samples for testing*

who has drunk a lot. They should also be able to identify which part of the nephron is malfunctioning if protein appears in the urine (glomerulus). If you have already taught pupils about diabetes, they will recognise the symptom of sugar in the urine.

References

Johnson, G. (1999) Kidney role-plays. *School Science Review* **80**:93–99

Newsome, J. & Lock, R. (1997) Modelling the urinary system. *Journal of Biological Education* **31**:256–257

chapter 14

Health

BACKGROUND

Human health depends on how effectively the body defends itself against
microbes (such as bacteria and viruses) which cause disease, and whether its
normal functions are disrupted by abuse of alcohol, tobacco, solvents or other
drugs. The body's defences include a layer of skin which encloses the body; any
wound in the skin is immediately sealed by the clotting of blood. If microbes are
ingested on food, they are destroyed by the acidic gastric juice in the stomach. If
they are inhaled, sticky mucus in the trachea (wind-pipe) catches them. Small
hairs (cilia) push the mucus into the back of the throat where it is swallowed. If
microbes evade these defence mechanisms, they are killed by white blood cells
(lymphocytes and phagocytes). Drugs, which disrupt the body's normal
behaviour, may be either depressant (suppressing the activity of the nervous
system) or stimulant (increasing the activity of the nervous system). Many drugs,
apart from their effects on the nervous system, also have side-effects which can
damage health.

Research suggests that even if pupils understand health education in school,
they do not relate that education to their everyday life. Pupils arriving from KS2
will know of some detrimental effects of alcohol, tobacco and other drugs on
health. They are unlikely to understand how those effects come about. At KS3,
pupils must be familiar with the effects of alcohol and other drugs on the body.
They must also be familiar with bacteria and viruses, their transmission and their
effects on the body, and how the body's defence systems deal with infection. At
KS4 double and single award, pupils must also be familiar with the effect of
solvents on the body, and understand in more detail the role of the defence
mechanisms.

Drugs

Pupils may have prior knowledge about drugs from PSE, and their own experience. Read your PSE curriculum before beginning this topic. Bear in mind that some pupils may know more than you about the effects of drugs. Instead of trying to push home a message of 'say no', it is better to approach drugs simply as another topic in the National Curriculum. This avoids giving pupils the impression that you are yet another adult purveying a moral message. As a result, pupils will respect the information gained in lessons, and will be more likely to make an informed choice about drugs. Make clear to pupils that they should not recount their own experience. If they tell you they have taken drugs or are in possession of drugs, you would be obliged to inform a member of senior management.

Pupils may ask if you have ever taken drugs. There are perils in answering yes or no. If you answer yes, you may reinforce their willingness to try drugs, and leave yourself open to criticism by parents and senior management. If you answer no, they may not see you as being suitably qualified to teach them. Your best policy is not to answer, and to explain why.

Ask pupils to define a drug: a chemical which affects a human's behaviour and physiology. Discuss how alcohol and paracetamol meet this description. Ensure pupils understand that drugs can have positive effects on health when taken in the right quantities.

Both types of drug may have side-effects. If any drug is taken in excessive quantities, the negative effects can have a serious influence on human health. The Health Education Council produces a range of leaflets about drugs. You may find some good videos amongst those normally used for PSE. Ask pupils to conduct research into the effects of drugs on the body. Examples include caffeine, nicotine, heroin, barbiturates, cocaine, LSD and cannabis. When complete, distinguish between the positive and negative effects of each.

Whilst doing this, they are likely to meet the word addiction. Most pupils will have some idea of what it means: if someone is addicted to a drug, they start to feel bad if they do not take it. Some pupils will mention physical and psychological addiction. If they do, you can discuss them (see KS4 below).

Alcohol and solvents are mentioned specifically in the National Curriculum. Spend some time concentrating on their effects. Pupils will be more familiar with alcohol and you can brainstorm their background knowledge. Define solvents as chemicals which dissolve other chemicals. Ensure pupils understand that people take solvents in vapour form, i.e. they sniff them. Again, the Health Education Council leaflets and other resources for research are likely to be available. Ensure pupils understand the positive and negative effects of alcohol and solvents, how they are taken, and what parts of the body they damage.

> ### Assessing pupils' learning
>
> - Ask pupils to present their research on the effects of drugs, including alcohol and solvents.
> - Ask pupils to explain why people must not drink and drive.
> - Set up a debate about whether drinking and driving should be legal, and whether cannabis should be legalised. Pupils could take the roles of doctor (who can explain the physical effect on the body), civil rights lawyer (everyone has a right to do what they want to), police (who can recount the legal aspects, and the aftermath of road accidents) and social worker (who has seen the division within families caused by alcohol and drugs). You could provide information sheets for each character, or get them to research the roles themselves.

KEY STAGE 4 CONCEPTS

Drugs

Again, begin by defining a drug as a substance which changes body functions. Distinguish between stimulants and depressants: these increase (stimulants) or decrease (depressants) the activity of the brain or nervous system, by speeding up or slowing down the speed of transmission of nervous impulses. More able pupils should understand how these drugs affect the nervous system. They act at synapses and affect the neurotransmitters (see Chapter 11).

- Nicotine (stimulant) copies the action of neurotransmitter molecules and triggers a nervous impulse in the post-synaptic neurone.
- Caffeine (stimulant) causes more neurotransmitter to be released from the pre-synaptic membrane.
- Alcohol (depressant) stops neurotransmitters reaching the post-synaptic membrane.

Ask pupils to research the medical, positive and negative effects of different drugs on the body, labelling each as a stimulant or depressant. Include a broader range than at KS3, including analgesics, anaesthetics, tranquillisers, hallucinogens and common stimulants. Demonstrate the effects of stimulants and depressants on water fleas (*Daphnia*) using the apparatus in Figure 14.1. Water fleas are likely to be available from your nearest garden centre. The cotton wool holds the water flea in one place to make it easier to observe. Count the

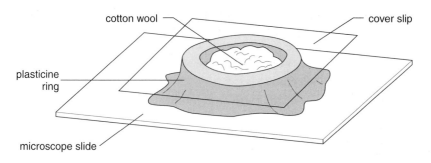

Figure 14.1 *Apparatus for looking at the heart beat of water fleas. Using a teat pipette, add one water flea to the cotton wool in the ring of plasticine. Examine under the microscope*

number of heart beats per minute in different concentrations of caffeine (stimulant) or alcohol (depressant). You can use this as an Sc1 investigation for more able pupils.

Pupils also need to be familiar with addiction at KS4.

- Distinguish between psychological addiction (the feeling that you can't live without a drug) and physical addiction (when your body begins to need a drug to work properly).
- Introduce the idea of tolerance: i.e. if someone is physically addicted, they will need more and more of the drug to have the same effect as when it was first taken.
- Ask pupils what happens when someone addicted tries to give up a drug: withdrawal symptoms, including nausea, vomiting, shaking and hallucinations.

Alcohol, solvents and tobacco are mentioned specifically in the National Curriculum. Because pupils normally show a wide interest in this topic, ask them to research the effects of these drugs, and to present that research to the class. Alternatively, give pupils information sheets about each type of drug, and set up a debate about whether each type of drug should be legal. The biology or PSE departments may have some good videos.

Syllabuses tend to differ in the detail required by pupils: examine your syllabus and past papers to assess this. Pupils should be aware that tobacco causes bronchitis, emphysema, lung cancer and heart disease. You can remind pupils of their work in KS3 and demonstrate the smoking machine again (see Chapter 7). The British Heart Foundation produces leaflets about the effect of smoking and alcohol on the heart. Your department may have prepared slides showing the effects of drugs on the lungs and the liver.

KEY STAGE 3 CONCEPTS

Spreading disease: bacteria and viruses

Begin by showing pupils the prevalence of bacteria around them. Ask your technician for some nutrient agar plates (Petri dishes containing nutrient agar). Pupils can culture bacteria from a variety of sources, including dust on the tops of cupboards etc. **Do not** collect bacteria from toilets, sinks, drains, skin, spots, pimples, noses, mouths, changing rooms, or any source where pathogenic bacteria may be located. Use the method in Figure 14.2 to plate out cultures; put lids on the plates and tape them closed so pupils cannot open them. Incubate them, upside down, at *room temperature* for about one week. You will see colonies of bacteria have grown on the agar. Do not open the plates again, and autoclave before disposal.

S **Safety Advice:** The precautions required for microbiological work are too extensive to be summarised properly here. Your department is likely to have *Safety in the School Laboratory* (ASE 1996) and/or *Microbiology: An HMI Guide*

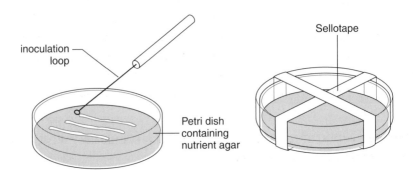

Figure 14.2 *Apparatus for culturing micro-organisms. Ensure you inoculate the plate in criss-cross straight lines, tape the plates closed as shown and incubate upside-down*

for Schools and Further Education (DES 1993). You must be familiar with the advice in these publications before beginning any microbiological work. Your technician and head of department will be able to advise.

To show that bacteria are in the air, put a little cotton wool (soaked in water) into a film case and place a couple of frozen peas on top. After a week at room temperature, or incubated at 25°C if possible, the colonies of bacteria (and fungi) will be apparent on their surfaces. If colonies have formed, spores must have floated through the air and onto the food.

Safety Advice: Warn pupils not to examine colonies closely. They should hold them at arm's length to reduce the chance of inhaling spores. Do not incubate above 30°C.

Conclude that bacteria are present throughout the environment. Remind pupils that viruses can also cause disease. Most pupils will not understand the difference between bacteria and viruses and will group them together as 'germs' or 'bugs'. Because the word bug is often used for insects, they may also link the two types of bug in their minds. More able pupils could do a research project on the structure of bacteria and viruses and their mode of infection. The main difference is that viruses reproduce inside cells and destroy them when they burst out. Bacteria normally reproduce outside cells but may digest and absorb cells, or may secrete toxins which disrupt normal cellular functions. Viruses are always much smaller than bacteria.

Most pupils will know some ways in which disease is spread. Establish the depth of their understanding by asking them to produce a booklet describing ways to avoid infection. Provide students with a list of sources of infection. Ask them to explain how each source could spread disease, and to think of ways in which spread of disease could be prevented. Examples include sneezing, going to the toilet, sexual intercourse, not cooking food properly, not storing food properly, sewage effluent being pumped into rivers, and open cuts.

Demonstrate the importance of storing food in the fridge by setting up film cases containing damp cotton wool, upon which are a few peas. Few bacteria and fungi will grow on those kept in the fridge. Conclude that low temperatures prevent microbes from reproducing and therefore reduce the number of microbes being ingested and the risk of them successfully surviving and reproducing inside the body.

Safety Advice: Be aware of the advice given in the publications detailed earlier before starting microbiological work.

113

KEY STAGE 4 CONCEPTS

Pupils do not need to focus on sources of infection. However, remind them that infection with bacteria and viruses can occur and group possible modes of infection into categories: microbes can be ingested, inhaled, taken in through cuts, or taken in during sexual intercourse.

KEY STAGE 3 CONCEPTS

Keeping the microbes out: the skin, blood, stomach and trachea

Briefly mention that the skin provides an impermeable barrier to almost all microbes. Ask pupils to think about regions of the body which lack skin, and to consider how they defend themselves against microbes: (i) ears have hairs and wax, (ii) the nose has hairs and mucus, and (iii) eyes have hairs (eyelashes), eyelids and tears to trap, kill or wash away microbes.

KEY STAGE 4 CONCEPTS

Keeping microbes out

Depending on your syllabus, pupils may need to know the detailed structure of the skin. If you have already considered it in Chapter 13 or Chapter 11, remind pupils of the structure and that the epidermis forms a continuous, water-proof barrier which prevents entry of microbes.

Some syllabuses require detailed knowledge of the body's response if the skin is punctured. Ask pupils what happens after they scratch themselves: it bleeds, then bleeding stops, a scab forms, and when healing is complete, the scab peels off. Foundation level pupils should know that platelets in the blood clog up the puncture and form a scab. Higher level pupils should be aware that puncture of the skin makes platelets break apart and release an enzyme called thrombokinase which acts on a protein prothrombin in the blood plasma and changes it to thrombin. This converts fibrinogen (a soluble protein in the blood plasma) to fibrin (an insoluble protein) which forms a net of fibres across the cut. This net traps red blood cells, preventing them getting out, and stops microbes from entering.

Ask pupils where there are gaps in the skin: (i) the mouth, giving access to the trachea, lungs and digestive system, (ii) the eyes, and (iii) the ears. Remind pupils of work from KS3 about eyes, ears and the nose. Tears and eyelids are designed to wash microbes away. Tears also contain an enzyme called lysozyme which kills bacteria.

To kill any microbes ingested on food, the stomach secretes hydrochloric acid. Remind pupils how sour their mouths taste after vomiting, i.e. when the contents of the stomach are ejected through the mouth. Remind them also that vinegar and lemon juice (both acids) taste sour.

To prevent infection upon inhalation, the nose is lined with hairs, and the trachea is lined by cilia (tiny hairs) and mucus. Remind pupils that when they have a heavy cough, they cough up mucus from their trachea. Your department may have prepared slides of ciliated epithelial cells (the type of cell in the trachea which bear the hairs) which pupils can examine and draw. The mucus is sticky and traps microbes and small dust particles. This is then swept up into the throat by the cilia, swallowed, and any bacteria are killed by stomach acid.

Assessing pupils' learning

- Less able pupils could produce a poster of the body, and annotate it with descriptions of defence mechanisms in each location.
- Ask pupils to write a curriculum vitae for the position of 'human invader', perfectly adapted to avoiding the body's defences.

KEY STAGE 3 CONCEPTS

Dealing with microbes if they get inside the body: the blood

Your department may have prepared slides of blood showing the two types of white blood cell: phagocytes (*phago* = eating, *cyte* = cell) and lymphocytes. Phagocytes kill microbes by 'eating' them (they engulf them and break them down with enzymes). At a cut in the skin, phagocytes squeeze out of capillaries and effectively 'walk' around the cut, hunting for microbes to engulf.

Tell pupils it is easier to find and 'hoover up' microbes if they are stuck together in big groups. Lymphocytes produce antibodies which stick the microbes together. You must provide pupils at KS3 with a very simple model of how antibodies are produced.

- Role play the action of antibodies: a group of people holding hands (microbes linked by antibodies) will find it more difficult to evade someone chasing them (the phagocyte) than individual people.
- Tell pupils about Edward Jenner (see most biology textbooks) and use his work as the basis for a play.

Pupils should understand that antibodies produced by lymphocytes must be specific to a particular microbe, and that when a new microbe enters the blood, the lymphocytes must learn how to produce antibodies to it.

Many pupils will know you cannot catch the same disease twice. All pupils should have been vaccinated against a number of diseases. If a new microbe enters the blood, it normally takes time for lymphocytes to make antibodies to it. If the lymphocytes have already encountered a microbe, antibodies can be produced so quickly that you may not even notice you have been reinfected: a person is therefore immune to a disease. By injecting a harmless version of the microbe, you can also give the lymphocytes practice in making antibodies.

Assessing pupils' learning

- Ask pupils to explain why they are vaccinated against diseases.
- Pupils could produce a role play or cartoon describing the action of phagocytes and lymphocytes.
- Pupils could write in the first person (i) as if they were a white blood cell, describing how they defend the body against infection, or (ii) as if they were a microbe successfully invading the body.
- Ask pupils to make a game showing how antibodies defend the body.

KEY STAGE 4 CONCEPTS

The role of the blood in defence

You can demonstrate the action of phagocytes to KS4 pupils using yeast and a microbe called *Paramecium*. *Paramecium* are commonly studied at A-level. *Paramecium* will engulf yeast cells in the same way as phagocytes engulf bacteria in the body.

- Stain a yeast cell suspension by adding a few drops of Congo red.
- Mix one drop of yeast cell suspension, one drop of methyl cellulose and one drop of *Paramecium* culture and observe on a microscope slide.

- If you watch long enough, you can observe the yeast cells being engulfed and breaking down inside the *Paramecium.*

The function and production of antibodies at KS4 is more complex. Again put the action of antibodies into context: remind pupils that you cannot contract the same disease twice and that a different antibody is needed for each different microbe. Pupils should understand how antibodies can affect invading microbes.

- Antibodies can clump microbes together to make them easier to phagocytose.
- Antibodies can puncture microbes and kill them.
- Antibodies can bind to toxic chemicals released by the microbes.

Immunity is more complex at KS4. Particular lymphocytes make antibodies to particular microbes; the antibodies are actually specific to antigens (molecules on the surface of the microbe). Suggest to pupils that if there were more lymphocytes who could produce antibodies to a particular microbe, then microbes would be eradicated more quickly. To achieve this, lymphocytes divide upon infection. Some cells (memory cells) stay in the blood ready to deal with the same infection again if the same microbe enters the blood at a later date. They normally deal with the microbe before it causes symptoms of the disease: hence you cannot catch the same disease twice. This is natural immunity.

   Immunity can be triggered without a person having to be infected with a live microbe. We can inject a killed or harmless form of a microbe, or an inactivated toxin from the microbe. These will trigger production of memory cells. This is called artificial immunity. Some pupils may ask why people can catch cold more than once: there are many different strains of cold virus and you cannot catch the same strain twice.

Assessing pupils' learning

- Pupils should write up the *Paramecium* experiment, explaining how the *Paramecium* engulfs the yeast, and drawing a parallel between this and phagocytes engulfing bacteria.
- Pupils should draw a cartoon to show how white blood cells defend the body.
- Pupils could write in the first person (i) as if they were a white blood cell, describing how they defend the body against infection, or (ii) as if they were a microbe successfully invading the body.

3 Green Plants as Organisms

chapter 15

Nutrition

BACKGROUND

Plants make their own food using energy from the Sun. This process is called photosynthesis: it uses light, carbon dioxide, water and chlorophyll (the green pigment in the plant's leaves) and produces glucose and oxygen. The word and symbol equations for this reaction are given.

$$\text{Carbon dioxide} + \text{Water} \rightarrow \text{Glucose} + \text{Oxygen}$$
$$6CO_2 \quad + \ 6H_2O \rightarrow C_6H_{12}O_6 + \quad 6O_2$$

The plant uses some of the glucose in respiration. Some is converted to starch (for food storage), cellulose (for cell walls), amino acids (for enzymes and structural proteins), fats and oils. Plant leaves are adapted to obtain the factors required for photosynthesis: they (i) are thin, allowing light to reach all photosynthesising cells and helping gaseous exchange, (ii) have a large surface area to volume ratio to ensure efficient absorption of carbon dioxide and release of oxygen (gaseous exchange), (iii) are broad to intercept a maximum amount of light, and (iv) have a dense network of veins, supplying water to photosynthesising cells. However, photosynthesis can only occur as fast as carbon dioxide, water and light are supplied to the leaves, and as fast as the chlorophyll molecules within the leaf can trap light energy and convert it to chemical energy. Whichever of these factors is in shortest supply is called the limiting factor; it limits the rate of photosynthesis. Plants also take in mineral salts, through their roots, to provide other elements required for life. The roots are adapted to maximising water and mineral salt uptake because they have a large surface area made up of thousands of single-celled root hairs protruding from the root surface.

All pupils should know from KS2 that plants use light to make food for growth, and that they need water. They should also know that plants take in some nutrients through the roots. At KS3, it is important to put plant nutrition into context. You could compare the nutritional processes of plants and animals, although some research suggests that this approach leads to confusion, because of the very different ways in which plants obtain their food (see below). At KS3 and 4 double award pupils should know the reactants and products of photosynthesis, the adaptations of the plant to obtaining the reactants for photosynthesis, the function of some minerals in the plant and the plant's adaptations for obtaining minerals. At KS4, they should understand the effect of limiting factors and the detailed role of minerals. Green plants are not included at KS4 single award.

KEY STAGE 3 CONCEPTS

The reactants required for photosynthesis

Ask pupils where plants get their food. Research suggests that most pupils think plants get food through the roots. This is very intuitive: they already know that humans take in their food from outside, and many will know that plants take up water through the roots. They may also know that plants are given 'food' by addition of fertilisers and that these are absorbed through the roots. Many pupils may believe that food *must* be ingested.

To overcome this, remind pupils that food supplies a respiratory substrate (glucose) for energy. If the plant makes its own respiratory substrate using light energy, this is still food, even if it has not been ingested. Unfortunately, many pupils believe that plants photosynthesise and do not respire, or that plants only respire at night. This is wrong: plants respire using the food made by photosynthesis. Animals respire using the food that they eat.

Suggest to pupils that if roots are responsible for getting the food required by a plant, it should not matter if you cover up the leaves. Use a plant which has been kept in the dark without water for a week and had its leaves wrapped in silver foil. Tell pupils that you have kept it well watered with rain water and its leaves covered in foil. When you take the foil off, the plant will look pale and wilted.

At this point, suggest that leaves are involved in making food for the plant by photosynthesis. Pupils may find it helpful to understand the derivation of the word photosynthesis. Most pupils are able to associate *photo* with light (e.g.

photography). You may need to explain that *synthesis* means 'to make'. Therefore, photosynthesis is 'making' food using 'light'. Tell pupils that the food made by photosynthesis is starch.

Ask pupils to confirm that plants make their own food by testing a leaf for starch:

- Dip a leaf in boiling water for 30 seconds.
- Cover with ethanol in a boiling tube (leave the stalk pointing upwards).
- Heat in an 80°C water bath until ethanol has leached the chlorophyll out of the leaf. The chlorophyll must be removed to allow the colour change from the starch test to be visible.
- Remove from ethanol, dip in water, spread out on a Petri dish and add iodine solution.
- If starch is present, a blue/black colour will appear.

(S) Safety Advice: Ethanol liquid and vapour are both highly flammable. If you have access to an electric water bath, use it in preference to students setting up their own over a Bunsen burner. If students must set up water baths using naked flames, ensure their Bunsen burner is off before you give them ethanol. Always restrict access to the ethanol and ensure you dispense it.

From KS2, pupils will know that plants are green (because they contain chlorophyll), and that they need light and water to survive. They may also know that plants take in carbon dioxide. If the absence of light, carbon dioxide or chlorophyll stop starch being produced, then they must be required for photosynthesis. Pupils can test whether each of these factors is required for photosynthesis to occur. Geraniums (*Pelargonium*) produce good results in each case and are easy to obtain. You must set up the plants at least three days before the lesson to control production of starch effectively.

Is light needed for photosynthesis?
Pupils should cover leaves with stencils of black paper; within reason they can design their own stencils. Attach the paper to the leaf with paper clips (Figure 15.1). Pupils should predict that only those parts of the leaf which are exposed to light will produce starch. Be aware that if the apparatus is left for several days, starch will migrate away from the region with access to light and into the rest of the leaf. Place the plant in the dark for three or more days to prevent starch production, and to ensure the leaf has broken down any starch formed previously. If, about three hours before the lesson, you then place the plant under a bench lamp, the shape of the stencils should show up distinctly when tested for starch. More able pupils should explain how the unilluminated areas act as the control with which to compare those areas which did have light.

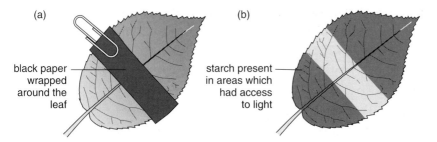

Figure 15.1 *Testing whether light is needed for photosynthesis. The leaf should be set up as shown a). When tested for starch, a black colour, indicative of starch, is formed in those areas with access to light b)*

Is carbon dioxide needed for photosynthesis?

Sodium hydroxide absorbs carbon dioxide from the air. This is caustic; pupils should not set this experiment up themselves, although they can test the leaves for the presence of starch. There are two sets of apparatus commonly used to deprive leaves of carbon dioxide; both are shown in Figure 15.2. Make the seal between bell jar and glass plate air-tight with vaseline. To ensure you obtain the results you expect (i.e. no starch), keep the plant in the dark for three or more days before the lesson.

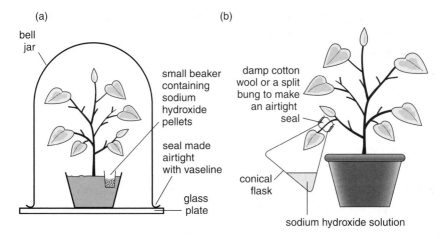

Figure 15.2 *Testing whether carbon dioxide is needed for photosynthesis. In both these apparatus, the sodium hydroxide absorbs carbon dioxide*

Is chlorophyll needed for photosynthesis?

It is not possible to remove chlorophyll from leaves and expect them to survive. For this reason, use a leaf from a variegated geranium which has cells containing chlorophyll in the centre and cells without chlorophyll around the outside. Again, keep these plants in the dark for three or more days, until three hours before the lesson. Then place them under a bench lamp. When tested for starch, pupils should predict that starch is formed only in those regions where chlorophyll is present.

Assessing pupils' learning

- Pupils should distinguish between the ways in which plants and animals obtain their food: animals eat it, and plants make it. They should also make clear that they understand that both plants and animals respire using this food.

- Pupils should write up experiments immediately upon completion. If they have all experiments to complete for homework, they tend to mix up results and methods. Less able pupils could simply draw a diagram for their method. For results, pupils may find it useful to draw their leaves before and after the experiment to compare the regions in which starch is produced. Conclusions should state whether each factor is required for photosynthesis.

- More able pupils should consider why it is impossible to test whether water is needed for photosynthesis (because water has so many functions, its absence would kill the plant).

- Give pupils data representing the fluctuations in carbon dioxide content of the air in an area containing vegetation during a 24-hour period. Pupils should explain that when there is light available, photosynthesis increases and carbon dioxide levels reduce. They could provide a similar explanation for the fluctuations in carbon dioxide levels over a year (there is more carbon dioxide in the air in winter because plants are not photosynthesising as quickly).

KEY STAGE 4 CONCEPTS

Much of the knowledge for KS4 is identical to that for KS3. Knowledge of the above practicals is likely to be examined in all GCSE examination board syllabuses.

KEY STAGE 3 CONCEPTS

The products of photosynthesis

Many students may know that plants produce oxygen; use this as a basis for
predicting that it is produced during photosynthesis.

Is oxygen produced by photosynthesis?

Use the apparatus in Figure 15.3. Canadian pond weed (*Elodea canadensis*) is
available from aquarium shops and garden centres. This is not an indigenous
species, so do not put spare pond weed into your local pond, as it smothers the
native species. You can probably put it in your department's fish tank. Pond
weed is used because it has a high photosynthetic rate. Because it is aquatic, the
oxygen produced can be seen as bubbles coming from the cut end of the plant's
stem. Tell the pupils you need to leave the experiment for a week to collect
enough oxygen. To test for the presence of oxygen in the test tube, lower in a
glowing splint and it should re-light. This test does require a large volume of
oxygen to work effectively. Because the pond weed is unlikely to produce enough
oxygen before your next lesson, add pure oxygen to your test tube from an
oxygen cylinder.

   If your department has data-logging equipment and an oxygen probe, you
could set up a beaker of pond weed and measure the oxygen content of the water
over a 24-hour period. During the day, the oxygen content should increase
because photosynthesis is happening.

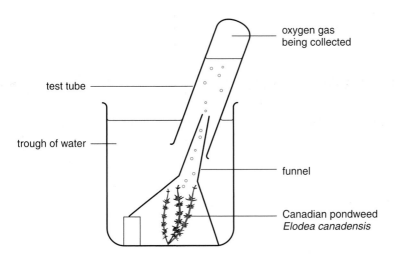

Figure 15.3 *Collecting the gas produced during photosynthesis*

Having completed these experiments, write a word equation:

$$\text{Carbon dioxide} + \text{Water} \xrightarrow{\textit{Light and Chlorophyll}} \text{Starch} + \text{Oxygen}$$

Tell more able pupils that photosynthesis actually produces glucose which is stored as starch by the plant. Note that light and chlorophyll appear above (or below) the arrow in the equation. Explain that the chlorophyll traps the light energy to allow the reaction to occur, but does not change chemically. Therefore, it should not be included as part of the chemical reaction. A symbol equation is not required at KS3.

Assessing pupils' learning

- Pupils should write up the oxygen experiment. Less able pupils could draw a diagram of a plant, showing where the reactants go into the plant and where the products emerge.
- More able pupils should explain the results of data-logging.
- Depending on whether they have met word equations in chemistry, more able pupils should be asked to write the word equation themselves. Give less able pupils boxes to cut out and rearrange correctly.

KEY STAGE 4 CONCEPTS

The products of photosynthesis

Students at this level should know that glucose is the product of photosynthesis. Starch comprises many smaller glucose molecules and is easier to store because it is insoluble and so does not affect water relations in the leaf's cells (see Chapter 18). Unlike starch, glucose is soluble and can be transported around the plant. You will need to change the word equation from KS3:

$$\text{Carbon dioxide} + \text{Water} \xrightarrow{\textit{Light} + \textit{Chlorophyll}} \text{Glucose} + \text{Oxygen}$$

More able pupils will also require the symbol equation. You can ask them to balance it:

$$6CO_2 + 6H_2O \xrightarrow{\textit{Light} + \textit{Chlorophyll}} C_6H_{12}O_6 + 6O_2$$

Limiting factors in photosynthesis

Pupils' understanding of this difficult topic will depend upon how you present the material. Start by re-stating that for photosynthesis to happen, all the 'ingredients' need to be there (carbon dioxide, water, light and chlorophyll).

Foundation level pupils will only need to realise that if there are inadequate ingredients, photosynthesis will slow down or stop. To demonstrate this, ask pupils whether photosynthesis happens when it is dark. Most pupils will correctly answer no.

More able pupils should understand that the factor (carbon dioxide, water, light or chlorophyll) in shortest supply will limit the rate of photosynthesis. For instance, even though there is plenty of water, carbon dioxide and chlorophyll during the night, photosynthesis cannot take place because of the lack of light. If you increase the amount of light slowly, the speed at which photosynthesis occurs will gradually increase.

Use a simple analogy to help understanding. Consider the case where chlorophyll is the limiting factor. Imagine the leaf is a factory which makes glucose. The chlorophyll represents the production line. The carbon dioxide, light and water are the raw materials. As you increase the supply rate of the raw materials, you can make the products more and more quickly. Eventually, the production line is working as fast as it can. It doesn't matter how much more quickly you supply raw materials, it cannot produce glucose any faster. The rate of photosynthesis is therefore limited by the rate at which the chlorophyll (the production line) can trap light energy and convert it to chemical energy in the glucose. You can extend this analogy to ask what happens when the production line is working as fast as possible, and suddenly the supply of carbon dioxide falls. The rate of photosynthesis clearly falls. The rate of photosynthesis has been limited by the availability of the carbon dioxide.

Demonstrating limiting factors

The speed of photosynthesis can be measured by monitoring the rate of oxygen production in Canadian pond weed. This could form the basis of an Sc1 investigation on the effect of light intensity on the rate of photosynthesis. Pupils should set up the apparatus in Figure 15.4 and count the number of bubbles of oxygen given off by the pond weed in a fixed period of time.

By moving the lamp closer to the plant, bubble production will increase up to a maximum (Figure 15.5). Pupils should explain why this happens in terms of limiting factors. Very leafy pond weed will give better results. You may need to add a few grams of sodium hydrogen carbonate to the water to ensure a high concentration of dissolved carbon dioxide.

More able pupils should understand that temperature can limit the rate of photosynthesis. As in any chemical reaction, photosynthesis will occur more slowly if it is cold, and more quickly if it is hot. There are two reasons for this: (i) simple kinetics – there are less collisions between reacting particles, and (ii) enzymes do not work as efficiently at the wrong temperature (see Chapter 6).

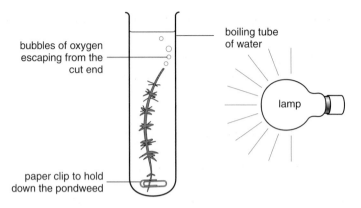

Figure 15.4 *Investigating the effect of light intensity on the rate of photosynthesis. Carry out the experiment in a dark room. Change light intensity by changing the distance between the lamp and the boiling tube*

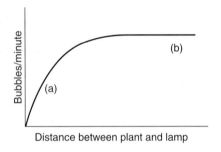

Figure 15.5 *Production of oxygen bubbles in response to light intensity*

You can demonstrate the effect of temperature by measuring the speed at which oxygen is given off when the boiling tube is placed into water baths of different temperature. Use a range of water bath temperatures between 10°C and 25°C. Because the plant will seal up its cut end, the bubbles will slow down over time. To gain the results you expect, begin the experiment at high temperature and then gradually reduce the temperature.

Assessing pupils' learning

- Less able pupils can complete simple questions about what happens when different factors are missing or in short supply.
- More able pupils should plot graphs of photosynthetic rate (represented by the speed of bubble production) against light intensity (or any factor which could limit the rate of photosynthesis). Ask pupils to identify which factor is limiting at different points on the graph (Figure 15.5). For

instance, at point 'a', light is limiting, and at point 'b' carbon dioxide may
be limiting.
- Ask more able pupils to write up the experiment changing temperature,
  explaining how temperature affects the rate of photosynthesis.

KEY STAGE 3 CONCEPTS

Adaptations of the leaf to photosynthesis

Pupils should be able to identify adaptations which help the plants gain access to
the factors required for photosynthesis. You could compare the man-made
structure of a solar cell, which is designed for maximum light interception, with
a leaf. Pupils should know that the leaf is the main site of photosynthesis and
that leaves are:

- green – because chlorophyll is needed for photosynthesis,
- thin and relatively large – to provide a large surface area for interception of
  light and gaseous exchange,
- supplied with a network of veins – to carry water from the roots.

Pupils should examine prepared slides of a transverse section of a
dicotyledonous leaf (your department is likely to have these). Most biology
textbooks contain diagrams of leaf sections. More able pupils could draw and
label from prepared slides; less able pupils should be given a diagram to label.

Point out the stomata (singular = stoma) – the pores in the lower surface of
the leaf which allow carbon dioxide to pass into the leaf and oxygen to leave. The
stomata lead to a network of air spaces between the cells inside the leaf. Pupils
should also be aware that the mesophyll layers of cells contain chlorophyll and
are therefore the site for photosynthesis.

Assessing pupils' learning

Pupils could

- write about the basic adaptations of the leaf to photosynthesis.
- make a model of the leaf cross-section using plastic (for the cuticle) and
  polystyrene blocks (for cells).

KEY STAGE 4 CONCEPTS

Adaptations of the leaf to photosynthesis

Pupils should be reminded of the material from KS3. They should also consider how the fine structure of the leaf is adapted to gaining the factors required for photosynthesis:

- Leaves have a thin and transparent epidermis which protects the leaf but also allows sunlight through to the photosynthesising cells, the mesophyll cells.
- The palisade and spongy mesophyll cells contain green organelles called chloroplasts, which are the sites of photosynthesis and can be seen easily under the light microscope.
- The palisade mesophyll cells contain the most chloroplasts, are columnar in shape and tightly packed together. This maximises the number of chloroplasts per unit area, and allows maximum interception of light for photosynthesis.
- The air spaces in the spongy mesophyll layer lead to the stomata in the lower leaf epidermis and allow gaseous exchange to occur. The extra surface area open to the air spaces provides a greater surface area across which gaseous exchange can occur.

128

Assessing pupils' learning

- Pupils should annotate a diagram, noting the function of the following structures: stomata, palisade mesophyll layer, air spaces in the spongy mesophyll layer, each epidermis and the waxy cuticle.
- Ask pupils to compare the adaptations of plants to those of animals for acquiring glucose and oxygen.
- Ask pupils to design a man-made machine, perfectly adapted to gaining the raw materials for photosynthesis.

KEY STAGE 3 CONCEPTS

What happens to the glucose produced by photosynthesis?

Pupils must know that some of the glucose produced by photosynthesis is used by the plant to provide energy through respiration. It is important that pupils realise that plants do respire. A common misconception is that plants 'do' photosynthesis whereas animals 'do' respiration. Research suggests many pupils cannot distinguish between the functions of photosynthesis and respiration: they label both as 'providing energy'. Remember: photosynthesis is the process by which plants capture light energy and turn it into a store of chemical energy (glucose). Respiration is the process by which that store of chemical energy is converted to a usable form of chemical energy (ATP) (see Chapter 5).

Although plants do photosynthesise *and* respire, respiration does not use up all of the glucose produced by photosynthesis; at the end of each day, there is a net gain of glucose. Consider deciduous trees (ones that lose their leaves in the winter); they can only photosynthesise when they are in leaf, but they manage to produce enough glucose for the winter months when they have no leaves and are not photosynthesising.

Assessing pupils' learning

- Provide pupils with data showing carbon dioxide excretion from a plant over 24 hours. They should explain (i) why carbon dioxide is produced at all by the plant (because the plant is respiring), and (ii) why almost none is excreted from the plant during daylight (because photosynthesis is using it during daylight).
- Pupils could produce a concept map, explaining the relationship between respiration and photosynthesis.

KEY STAGE 4 CONCEPTS

What happens to the glucose produced by photosynthesis?

Remind pupils of the material from KS3. Confusion between the processes of photosynthesis and respiration often remains, even in some more able pupils, through to KS4. KS4 pupils should know the fate of glucose in more detail. This would be a good opportunity for research from textbooks or CD-ROMs.

- Glucose is used immediately to provide energy through respiration.
- Some is stored as sucrose (energy store in fruit) and starch (energy store in leaves, seeds, roots and tubers).
- Glucose is a raw material for growth. Research suggests that pupils fail to realise the importance of photosynthesis to growth, and sometimes suggest that minerals taken in from the roots are used for growth, whereas glucose made in the leaves is exclusively used for energy. In fact, a large amount of glucose is used to make cellulose, the main structural molecule in cell walls.
- The rest of the glucose is used to make oils (energy store in seeds), proteins (enzymes and cell wall proteins), DNA, hormones and indeed all other plant molecules.

Assessing pupils' learning

- Pupils could produce a booklet or leaflet about the fate of glucose.
- More able pupils could be encouraged to research the chemical reactions involved in changing glucose to other compounds.

KEY STAGE 3 CONCEPTS

Mineral nutrition in plants

As outlined above, photosynthesis provides an energy source (glucose) which is made up of the elements carbon, oxygen and hydrogen. However, plants need other elements to grow and function properly. The only source for these elements is absorption. Minerals are absorbed through root hairs and are present either naturally in the soil, or from fertilisers. Mineral deficiency symptoms include yellowing, reddening or scorching of the leaves. Different symptoms are diagnostic of different deficiencies.

Cress, grown on cotton wool, is excellent for demonstrating the presence of root hairs. Pupils should know that minerals are absorbed across the root hairs and that the root hairs provide a larger surface area for absorption. To convince pupils of this, provide them with drawings of two roots, both 10 cm long, but with one having root hairs sticking out from the sides. Ask them to measure the length across which minerals could be absorbed in each root; they will need a string to measure the root with root hairs. Repeat the drainpipe demonstration from Chapter 6 to stress the importance of surface area.

Pupils should be familiar with the functions of a variety of minerals in plants. These should include nitrogen and a range of others. You could set up plants with a range of mineral deficiencies, as described below in KS4. Bear in mind that the results from such practicals may not duplicate the deficiency symptoms described in textbooks!

Assessing pupils' learning

Ask pupils to research the functions of different minerals and produce a booklet or table listing these functions in the plant.

KEY STAGE 4 CONCEPTS

Mineral nutrition in plants

Pupils should be aware that root hairs are single cells and that they increase the surface area available for absorption of mineral salts. They should also be able to label a simple diagram of a transverse section of a dicotyledonous root. Your department is likely to have prepared slides which more able pupils can draw. Give all pupils schematic diagrams (those found in textbooks) to label. Pupils should be aware of the function of a wide range of minerals in the plant, including nitrogen, phosphorus and potassium (N, P & K listed in fertilisers). Nitrogen increases leaf growth, phosphorus increases root growth and potassium increases flower and fruit production. Depending on the syllabus, they may also need the functions of iron, manganese, molybdenum, boron, zinc, copper, calcium, magnesium and sulphur. They should know the symptoms exhibited by the plant if each element is in short supply. More able pupils should be encouraged to know the *exact* function of certain minerals. For example, magnesium forms part of the chlorophyll molecule; nitrogen forms part of amino and nucleic acids.

Pupils can predict the effect of fertiliser (rich in minerals, particularly nitrogen, phosphorus and potassium) on plant growth.

- *Science and Plants for Schools* (SAPS: http://www-saps.plantsci.cam.ac.uk/) suggest how to use radishes to investigate the importance of fertiliser or different mineral salts (e.g. potassium nitrate, potassium chloride, calcium nitrate, magnesium sulphate and calcium chloride) on growth.
- Grow duckweed (*Lemna minor*) in different fertiliser concentrations. This grows very quickly and you can measure growth according to the number of leaves produced. Add a few grams of sodium hydrogen carbonate to ensure photosynthesis takes place. Bear in mind that too much fertiliser will be detrimental to growth. You may have to experiment before the lesson to decide what concentrations of fertiliser to use for best results.
- Grow mustard seeds on blotting paper soaked with solutions of different minerals. Compare fertiliser and no fertiliser, or vary the concentration of particular minerals to observe the effect on growth. Weigh the seeds before the experiment. After two weeks growth, dry the seedlings in an oven at 100°C and re-weigh. Calculate the percentage increase in mass in each case.

More able pupils should consider how minerals are absorbed by the root hairs. Because the concentration of minerals within the root hair cytoplasm is high, minerals tend to diffuse back out into the soil. However, because the minerals are so valuable to the plant, the root hairs actively pump the minerals back into the plant. It is important to stress that minerals are moving from a region of low concentration to a region of high concentration within the root and that energy must be used to do so. This process is called active transport (see Chapter 4).

Assessing pupils' learning

- Pupils should make a booklet or table showing the functions of minerals in the plant.
- More able pupils should write about active transport through structured questions.

chapter 16

Reproduction

BACKGROUND

Plants can reproduce sexually or asexually (see Chapter 10). The flowers contain the plant's sex organs. The male sex organs are the stamens, made up of pollen sacs (anthers) at the distal end of stalks (filaments). The female sex organs are the pistils (sometimes called carpels), comprising the stigma, style and ovary. The male parts of the flower are collectively known as the androecium; the female parts the gynaecium. The male sex cells are the pollen cells; the female sex cells the ova (eggs) which are contained within the ovary. Some flowers only have male parts, some only have female parts, and some have both. Self-pollination occurs when pollen cells are transferred from an anther on an individual plant to any stigma on the same plant. Cross-pollination occurs when pollen cells are transferred from an anther on one plant to a stigma on a different plant. Pollen may be blown to the stigma by the wind (especially in plants that produce catkins) or it may be transferred by an insect. When a pollen grain lands on a stigma, a pollen tube grows down the style into the ovary and the pollen cell passes along the tube and fuses with the ovum (fertilisation).

Pupils are familiar with plant reproduction from KS2. They should already understand that plants make seeds which are dispersed and germinate into new plants. At KS3, they must also know about sexual reproduction, how pollination and fertilisation occur and how seeds are formed. Plant reproduction is not included at KS4.

Types of reproduction

If you have already studied reproduction in humans, you may have discussed the differences between sexual and asexual reproduction (see Chapter 10). If so, remind pupils of the differences. Both types of reproduction take place in plants.

KEY STAGE 3 CONCEPTS

Asexual reproduction

When introducing asexual reproduction, you can demonstrate that plants can make identical offspring asexually (i.e. without involvement of sex cells).

- Take a stem cutting from a geranium (*Pelargonium*), *Begonia* or privet (*Ligustrum vulgare*) and place in a beaker of water. Once it has grown roots, plant it in compost: a new individual will take root. Unless you have a greenhouse, cuttings may not take root in the winter.
- Detach a leaf from any succulent plants which have easily detachable leaves, e.g. *Kalanchöe* and *Crassula*. If you lie the leaf on dry soil with the cut end in contact with the soil, the stalk will grow roots, forming a new independent plant.

Ask more able pupils to research the methods which plants use naturally to reproduce asexually:

- producing a runner (a shoot which grows horizontally above the ground) at the end of which a new individual plant grows, forms its own roots and breaks away (e.g. spider plant, strawberry).
- producing rhizomes (a shoot that grows horizontally below the ground) at the end of which a new individual plant forms its own roots, grows and breaks away (e.g. couch grass, mint).
- producing tubers (food stores are formed under the plant at several points along the stem or roots) from which new plants can grow (e.g. potato).
- producing bulbs which grow from lateral buds on old bulbs. Each new bulb can grow into a new plant (e.g. shallots, garlic).

> ### Assessing pupils' learning
>
> - Ask pupils to identify the differences between sexual and asexual reproduction.
> - Pupils should write up experiments, explaining that taking a cutting is a form of asexual reproduction because it does not involve sex cells.
> - More able pupils can identify different forms of asexual reproduction from example pictures, or present a poster about different methods of asexual reproduction.

Sexual reproduction

Pupils often fail to realise that for reproduction to be sexual, there must be fusion of gametes. Some will think there must be sexual intercourse, and will label plants asexual as a result. Most will not realise that the flower contains the reproductive organs. Revisit the human reproductive system. There are gametes (sperm and eggs), and there are sex organs which produce the gametes and get them into the same place for fertilisation to occur (male and female sex organs). Plants are exactly the same. The gametes are pollen cells and egg cells. The sex organs are contained within the flower.

Depending on the time of year, pupils can dissect and draw some simple flowers. *Primula* flowers have both male and female parts. If you cannot get *Primula*, cherry tree blossom is easy to get hold of, although is rather small for demonstration. In other species, some flowers will have only male parts, and some only female parts. If you want to use these, *Begonia* and shepherd's purse (*Capsella bursa-pastoris*) are good examples. The latter flowers through the year.

- Work through the flower, from the outside (starting at the sepals) to the inside (finishing at the stigma, style and ovary).
- Pupils should be able to name all the structures of the flower, and be aware of their functions (you will find them in most textbooks).
- More able pupils could examine the structure of stamens and pistils under a binocular dissecting microscope. They could also tap the anthers onto a microscope slide, add a drop of water and a cover slip and observe the pollen.
- Pupils can draw the flower but most such drawings tend to be of poor quality. All but the best students will need a textbook diagram to help interpret their flowers.

Having understood the sexual anatomy of plants, move on to pollination. The main difference in reproduction between plants and animals is the method by which the male gametes (pollen cells) are transferred from the anthers to the ovary of the recipient flower. Many pupils will know that people suffer from hay fever because of pollen in the air. Suggest that pollen is blown through the air from the anthers of some flowers to the stigmata of others (wind pollination). Pupils will also have seen bees feeding from the nectar of plants. The insects brush against the anthers, and carry pollen to the stigmata of other flowers (insect pollination).

- Give pupils two different flowers: one which is insect pollinated (e.g. honeysuckle (*Lonicera*) and buttercup (*Ranunculus*)), and one wind pollinated (e.g. most grasses and hazelnut). Ask them how each flower is adapted to its method of pollination. For example, insect pollinated plants are normally colourful, aromatic and produce nectar. Wind pollinated flowers do not need these adaptations, but have large anthers hanging in the wind and often a complicated feather-like stigma to trap wind-borne pollen.

- More able pupils could examine pollen from wind- and insect-pollinated plants under the microscope. Some pollen from insect-pollinated plants has tiny hooks to attach to the insect.
- *Science and Plants for Schools* (SAPS: http://www-saps.plantsci.cam.ac.uk/) describe the use of rapid cycling brassicas for studies of pollination. Plants will flower about 14 days after germination, and seeds are produced about 20 days after pollination. Pupils can pollinate their own plants, and watch the development of seeds.

Finally, deal with fertilisation. When a pollen grain arrives on the stigma, a tube grows down from it to the ovule within the ovary. The nucleus of the pollen cell then moves down the tube and fuses with the female nucleus. *Science and Plants for Schools* have a method to stimulate growth of a pollen tube artificially.

Some plants have methods to prevent self-fertilisation: if plants fertilise themselves, their offspring contain the same genes. If many genetically-similar individuals are produced, they will all be susceptible to the same diseases. Cross-fertilisation increases variability; people are not identical to either parent because they received genes from both. Once each ovule is fertilised, it begins to divide to form (eventually) a seed. Pupils could produce a leaflet or flow chart explaining how fertilisation occurs in plants.

A single ovary may contain one or more ovules (each containing an ovum), and therefore one or more seeds. An ovary containing fertilised ovules is called a fruit. For instance, a pod of beans is the fruit; the beans are the seeds. A tomato is a fruit containing individual seeds. More able pupils can consider how seeds are dispersed. The wall of the ovary often swells up and becomes sweet, brightly coloured and aromatic. A sweet fruit develops to attract animals. If animals eat fruit, they will digest the fleshy part, but the seeds have a tough resilient coat or testa which resists digestive enzymes in the intestine of the animal. Therefore, when the animal defaecates, the seeds are dispersed. Other seeds may be dispersed through the air. For instance, sycamore fruits, containing seeds, grow 'wings' and dandelion 'parachutes' may be carried along in the air.

You can do three practical things with seeds:

1 Pupils can study what happens during germination of bean seeds using the apparatus in Figure 16.1. Mung beans will germinate overnight in a warm place.

2 Pupils can dissect a pea (seed). Begin by finding the scar where the pea broke away from the fruit (the pod). Next to that scar is a tiny hole called the micropyle (the hole through which the pollen cell nucleus entered the ovule).

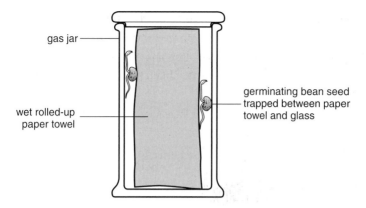

Figure 16.1 *Apparatus for germinating bean seeds. The diagram shows a cross-section across a gas jar containing a damp, rolled-up paper towel. Sandwich the beans between the towel and the glass*

If they remove the outer testa (the covering), they can split the remainder into two cotyledons. Ask pupils what they think the cotyledons contain (food stores). Between the cotyledons, you will notice the plumule which will grow into the shoot, and the radicle which will grow into the root.

3 Pupils can investigate what factors are needed by seeds for germination using the apparatus in Figure 16.2.

S **Safety Advice:** Alkaline pyrogallol is corrosive. If you would prefer not to use it, use iodine instead and kill the seeds placed into the pyrogallol tube by boiling for five minutes before the lesson (don't tell the pupils!).

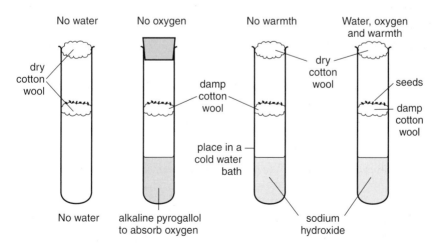

Figure 16.2 *Investigating whether water, oxygen or heat are needed for germination*

137

Science and Plants for Schools (SAPS: http://www-saps.plantsci.cam.ac.uk/) produces a template to make a model of a germinating seed from a sponge. The model is particularly good, because you can screw the sponge into a ball as it dries. When you add water, it behaves just like a seed with the active emergence of a plumule and a radicle.

Assessing pupils' learning

- Pupils can design flowers which are perfectly adapted to either wind pollination or insect pollination.
- Pupils can design seeds and fruits which are perfectly adapted to animal dispersal or wind dispersal.
- All pupils should show written understanding of the process of fertilisation. More able pupils could write about the process, or draw a flow chart. Less able pupils could complete a cloze exercise to describe the sequence of events involved in fertilisation.
- Pupils should draw or label diagrams of flower and seed anatomy, explaining the function of all their components.

chapter 17

Hormonal control

BACKGROUND

Unlike humans, plants do not have nerves. Almost every aspect of plant development is therefore controlled by hormones. Hormones control so-called tropic responses in plants: a directional response made by a plant to an external stimulus. This response may be to light (phototropism) or gravity (geotropism). For example, if a shoot grows towards the light, it is positively phototropic. If a shoot grows upwards, away from gravity, it is negatively geotropic. These responses are controlled by a hormone called auxin whose distribution in a shoot or root is affected by the external stimulus. Because auxins affect growth, if they are unevenly distributed within a shoot or root, they will cause uneven growth, and the shoot or root will bend in a particular direction. Hormones also control so-called nastic responses. These are non-directional responses to an external stimulus. For example, a tree sheds its leaves in response to day length; a venus fly-trap shuts its traps in response to the presence of an animal in the trap. Farmers sometimes apply hormones artificially to crops to speed up the onset of fruiting or germination.

Hormonal control of plant development is not included at KS2 or 3. At KS4 double award, it is useful to teach hormonal control in plants after having taught hormonal control in humans. At KS4 double award, pupils must be aware of the mechanism of hormone action and the commercial uses of plant hormones. Green plants are not included at KS4 single award.

KEY STAGE 4 CONCEPTS

What do hormones control?

Begin by asking pupils what nerves and hormones do in humans: they co-ordinate responses to external stimuli. Remind them of the distinctions between the two control systems: the nervous system gives fast responses and the hormonal system gives slow responses. Ask pupils if plants have nerves: most will reply correctly that they do not. In that case, plants must have hormones to control them.

Brainstorm aspects of a plant's life which are controlled by hormones: e.g. flowering, growing, fruiting, bending towards light, germination, dormancy in winter, opening and growth of buds. Distinguish two types of responses by plants and give examples of each:

- Nastic responses are non-directional responses to an external stimulus. For example, flowering in response to day length. You can buy venus fly-traps (*Dionaea muscipula*) at most garden centres. Demonstrate the traps closing (a nastic response). Do not do this too often or the plant will die. Sensitive plants (*Mimosa pudica*) also show a nastic response to touch; their leaves fold up.
- Tropic responses are directional responses to an external stimulus. For example, bending of the stem towards light. If a plant's response is towards the stimulus, the tropism is described as positive. If it is away from the stimulus, the tropism is described as negative. You should also define phototropism (a response to light) and geotropism (a response to gravity). Remind pupils that *photo-* is associated with light (e.g. in cameras) and *geo-* is associated with the earth (e.g. geography, geology). All pupils should remember that plants grown on the window sill bend towards the light. Also distinguish between *tropic* (a response to an external stimulus) and *trophic* (feeding; see Chapter 24) to avoid confusion. Pupils should understand why photo- and geotropisms are important to the survival of plants (light is used for photosynthesis; growing towards gravity normally gives a root access to water).

Assessing pupils' learning

- Ask pupils to define tropic and nastic responses (provide the definitions for less able pupils) and distinguish between photo- and geotropisms and positive and negative tropisms.
- Pupils should correctly identify such responses from a circus of situations, e.g. venus fly trap closing on an insect, growing towards light, flowering, leaves dropping.

Tropic responses

You can use practicals to demonstrate phototropisms (Figure 17.1a). Pupils could set them up during the lesson, or you could set them up as a demonstration. Ensure you give at least five days to obtain a convincing result. Pupils could extend the shoe box experiment by putting a sprouting potato at one end of the box. They can set up a maze using pieces of cardboard in the box to see if the potato can grow around the cardboard to reach the light (Figure 17.1b). Ensure that the only light shining into the box is from the hole at the end.

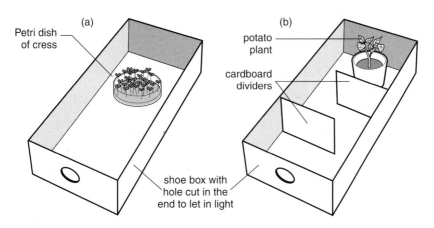

Figure 17.1 a) *Petri dish of cress contained within a shoe box. Because there is only one light source through the hole, the plants will bend towards the light, showing a positive phototropic response. b) Obstacle course around which the potato must grow to reach the light source*

Tell pupils that tropic responses are controlled by a hormone called auxin which causes growth. You can set up an experiment to show whereabouts in the shoot auxin is made (Figure 17.2). If the shoot tip is removed, no growth occurs. This experiment requires time to work: you could simply give pupils the method and results. Growth actually occurs just behind the tip in the region called the meristem. All cells in the tip produce auxin which diffuses back to the meristematic cells and promotes growth.

You are now ready to explain phototropism. Pupils find this quite difficult; try the following. On a piece of paper, ask pupils to draw around the cress shoots which have bent towards the light. They should measure the two sides of the shoot on their drawing. The outer edge is longer than the inner edge and therefore must have grown more quickly. So, if the shoot bends towards the light, the dark side must have grown more quickly than the light side. If the dark side grows more quickly it must have more auxin.

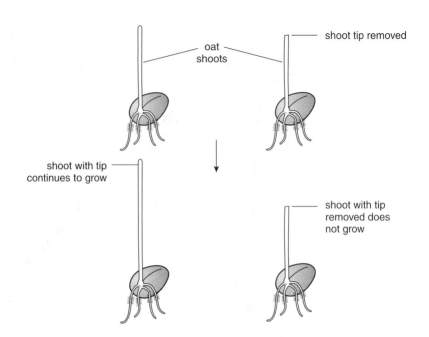

shoot tip removed

oat
shoots

shoot with tip
continues to grow

shoot with tip
removed does
not grow

Figure 17.2 *Apparatus to demonstrate that auxins are made in the growing tip*

Auxin actually diffuses to the dark side of the shoot away from the light, hence causing more growth. You may need to provide students with diagrams of how auxins diffuse in the shoot to help them understand (Figure 17.3). Ask more able pupils to predict the results of the experiments in Figures 17.4 and 17.5.

Having explained phototropisms, geotropisms should be easier. To demonstrate geotropisms in beans set up the apparatus in Figure 16.1 over seven days before you need it. The shoot emerges from the seed horizontally then turns and grows upwards: pupils should be able to predict which side is growing faster, and therefore that auxin is most concentrated on the lower side of the shoot. If you then turn the gas jar on its side, the shoot will change direction again. You can explain the collection of auxin on the lower side as being a result of gravity (Figure 17.3).

Ask pupils to explain why roots grow downwards. Many pupils will be confused: if auxin is pulled to the lower side of the root by gravity, that side should grow more and the root should bend upwards. However, they know that does not happen. The most able pupils may realise that auxin has a different effect in roots: it inhibits growth. This means that the upper side continues to grow normally, whereas the lower side stops growing.

For more able pupils confirm your explanation with a klinostat (Figure 17.6). Your department is likely to have one and advise on setting it up. Make sure you place the klinostat in the dark or the shoot will simply bend towards the light. In

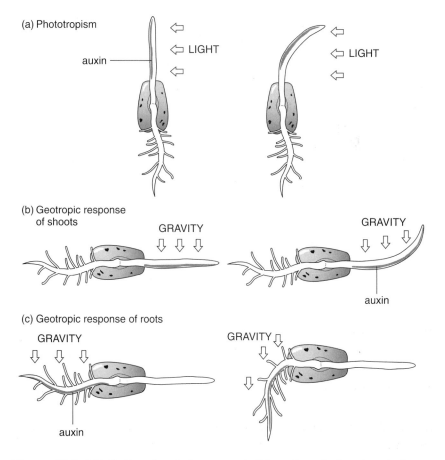

(a) Phototropism

auxin — LIGHT

LIGHT

(b) Geotropic response
of shoots

GRAVITY

GRAVITY

auxin

(c) Geotropic response of roots

GRAVITY

GRAVITY

auxin

Figure 17.3 *Distribution of auxin in response to light and gravity in a shoot and root,
and its effect on tropic responses*

fact, the shoot and root grow out straight. This is because the auxin is kept
moving around the plant as the klinostat rotates (because gravity keeps pulling
the auxin downwards); this means all of the shoot cells come into contact with
auxin at regular intervals and growth occurs evenly.

Assessing pupils' learning

- Pupils should demonstrate written understanding of how auxin controls
  tropic responses in plants.
- Ask pupils to predict the direction of growth of shoots and roots which
  have been manipulated by having their tips removed and: (i) replaced
  with normal agar, (ii) replaced with agar containing auxin, and (iii)
  replaced over half the cut end of the shoot with agar containing auxin.

- Ask pupils to predict how plants would grow in the zero-gravity environment of space (e.g. inside the space shuttle).
- Ask pupils to explain why the shoot and root grow straight when put on the klinostat.
- More able pupils could do research about the hormones involved in nastic responses in plants.

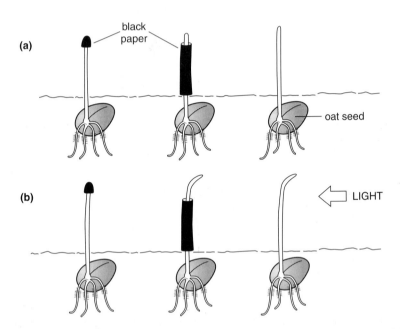

Figure 17.4 *Auxin is made in the growing tip and diffuses down to the dark side of the growing shoot. This causes the shoot to bend towards the light. When the tip is covered, there is no change in distribution of auxin and no phototropic response*

Industrial uses of plant hormones

Pupils should research how aspects of plant development could be manipulated for commercial ends (your department will have textbooks or leaflets).
- Growth hormone can stimulate cuttings to produce roots and form new plants.
- Harvested crops can be kept dormant and prevented from spoiling.
- Fruit ripening can be accelerated or delayed according to demand for the crop.
- Selective weed-killers can be used. These make weeds grow too fast. The weeds cannot sustain the fast rate of growth and they buckle and die.

You could give pupils data showing the difference in fruit yield between crops treated and untreated with hormones. If questions link crop yield to profit,

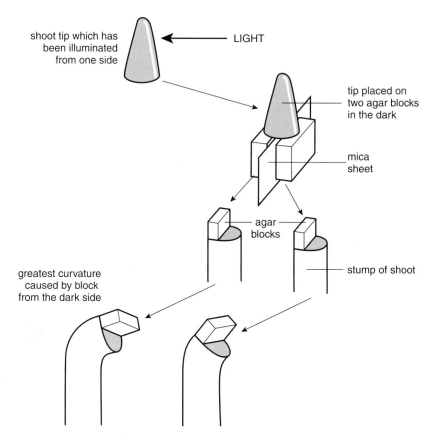

shoot tip which has
been illuminated
from one side

LIGHT

tip placed on
two agar blocks
in the dark

mica
sheet

agar
blocks

stump of shoot

greatest curvature
caused by block
from the dark side

Figure 17.5 *When the tip is removed and placed on the agar block, auxin diffuses from the tip into the agar. Because auxin diffuses away from the light, the agar block from the dark side will contain most auxin and cause most growth*

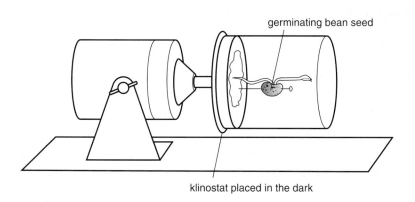

germinating bean seed

klinostat placed in the dark

Figure 17.6 *A klinostat set up to investigate geotropic responses. The shoot is rotated by the Klinostat*

they will come to understand how important hormones are in modern agriculture.

Assessing pupils' learning

- Pupils could make a booklet or poster showing the commercial uses of plant hormones.
- Pupils should answer questions based on data about the commercial importance of plant hormones to agricultural and horticultural yields.

chapter 18

Water and the plant

BACKGROUND

Plants need water for support, photosynthesis, cooling, and movement of minerals and organic compounds. Water is absorbed in the roots, moves up the stem into the leaves, and is eventually evaporated from air spaces within each leaf. Such evaporation is known as transpiration. Water is said to move through the plant by the transpiration stream. Water moves within plants by simple diffusion and osmosis (Chapter 4). Water and mineral salts are transported around the plant in the xylem vessels (long hollow tubes of dead tissue). Water and organic compounds are transported around the plant through phloem sieve tubes (long tubes of living cells arranged end-to-end and separated by perforated cell walls).

Pupils should know from KS2 that plants need water to grow and that water and minerals are taken into the plant through the roots. However, research reveals some serious misconceptions: for example, (i) water enters a plant through the leaves (bear in mind that when children water plants, they normally pour water onto the leaves), (ii) all the water absorbed by the plant is retained (i.e. none is lost through the leaves), (iii) water leaves the plant through the flowers, (iv) water leaves the leaves as a liquid, and (v) water is pumped around the plant in the same way as the heart pumps blood around the human body. The activities described in this chapter are aimed at overcoming these problems. In addition to the understanding gained at KS2, KS3 pupils must know that water is taken into the plant through the root hairs which provide a greater surface area for absorption. At KS4 double award, they must know that water is involved in transporting substances around the plant and that water is important in supporting the plant. Green plants are not included at KS4 single award.

KEY STAGE 4 CONCEPTS

The function of water and its effect on plant cells

Begin by recapping functions of water in the plant which are familiar: photosynthesis and cooling (just like sweating in humans). In the rest of this topic, you will introduce pupils to the structural and transport role of water. To understand the effect of water on plant cells, pupils must understand osmosis (see Chapter 4). The practice and theory of the onion experiment described in Chapter 4 will let pupils see the effect of water on plant cells. Use this practical to explain why cells become flaccid and turgid, in terms of relative water concentration. More able pupils should draw cells from the slide. Less able pupils could copy diagrams of turgid and flaccid cells from textbooks. Use the model in Figure 4.3 to demonstrate what happens when water enters the cell; the cell membrane pushes up against the cell wall. Point out that the cell wall prevents the cell membrane from expanding too far: the cell is turgid. When the balloon deflates (water moves out of the cell), the balloon pulls away from the box (the cell membrane pulls away from the cell wall) and the cell becomes flaccid.

Use the potato experiment in Chapter 4 to confirm the above. Ask pupils to explain what happens to the cells in the pieces of potato in each solution. The cells in the stiff piece of potato are turgid (full of water) and the cells in the floppy piece of potato are flaccid (not full of water).

Conclude this section by comparing the behaviour of animal cells in solutions of different water concentration. To avoid using cheek cells, you can use liver cells. A method to prepare these is described by Burton (1999). Begin by asking pupils what happens when a plant cell fills with water by osmosis. They will respond that the membrane is pushed up against the cell wall. Ask them what would happen in an animal cell without a cell wall: it would burst. The cell wall therefore gives structural reinforcement to plants.

Assessing pupils' learning

Ask pupils to

- write up the onion and potato experiments, explaining the difference in flaccid and turgid cells in terms of movement of water by osmosis.
- predict the direction of water flow, either into or out of the cell when placed in solutions of different concentrations.

KEY STAGE 3 CONCEPTS

Vascular tissue and the stem, roots and leaves

Begin by establishing whether pupils know where water enters the plant. You could do this with questions, e.g. where is the best place to give water to a plant: on the leaves or in the soil? If pupils do believe that water is taken in through the leaves, ask them to set up two plants: one which has its roots in water, and one which is uprooted and placed upside down into a vase so the leaves are submerged. They will see that the plant with its roots in water survives and the one with its leaves in water withers.

Pupils should know there are two types of tube which carry substances around the plant. Water and minerals are transported from the roots to the rest of the plant through the xylem (pronounced *zylem*) vessels. Sugar, made by photosynthesis, is transported from the leaves to the rest of the plant through the phloem. Cut a stick of celery, preferably still with some leaves on, and place it into a solution of red food colouring. Allow at least one hour for the red dye to move up the stem. Cut a section from the celery; you will see the location of the xylem vessels which now contain dye.

Your department should have some prepared slides of cross-sections of the root, stem and leaf, showing the positions of xylem and phloem. Pupils could draw these, or label prepared diagrams. Pupils should be aware of how the roots are adapted to absorbing water: they have root hairs (single cells which protrude from the root) which increase the surface area for absorption (see Chapter 15). You can see root hairs on the roots of cress germinated on tissue paper.

Assessing pupils' learning

- Pupils should label cross-sections of the stem, root and leaf.
- Less able pupils could split into groups and prepare models of the stem, root and leaf and present those models to the class. Polystyrene and straws are useful for model making.
- Ask more able pupils to explain how the presence of root hairs increases absorption of water.

KEY STAGE 4 CONCEPTS

Vascular tissue

Pupils must know all of the above, and be able to recognise the position of xylem and phloem in diagrams of the cross-section of the root, stem and leaf. Ensure you mention that xylem and phloem 'tubes' are collected together in vascular bundles which run through the plant. You can make a model of a vascular bundle using different coloured straws to help pupils understand that the bundle is made up of lots of individual tubes.

Many students find it difficult to appreciate how cross-sections of biological specimens relate to the actual specimens. Ensure you explain that cross-sections on slides were prepared by cutting a very thin section across the specimen.

At KS4, more able pupils must also recognise the structure of xylem vessels and phloem sieve tubes. You can find these in many textbooks. Stress that both tubes are made of cells arranged end to end. In the xylem, the intermediate cell walls have broken down, and a substance called lignin has been deposited which actually kills the cells. The xylem tubes are therefore dead. Phloem tubes are living, but the intermediate walls have only broken down partially to make 'sieve plates' between the cells through which sugar and water can pass.

Many syllabuses may require a detailed knowledge of root structure, specifically with regard to root hairs. Root hairs are single cells which increase the surface area available to the leaf for absorption of water and mineral salts. As mentioned above, you can see these on the roots of cress, or on prepared slides. Your department should have prepared slides of root sections.

Assessing pupils' learning

In addition to those activities mentioned for KS3 above:

- Pupils could make models (using lemonade bottles to represent cells) of xylem vessels and phloem sieve tubes. They should include cardboard sieve plates to separate the bottles in the phloem and paint them different colours to signify that the xylem are non-living and the phloem living.
- Pupils could make a table of similarities and differences in structure and function of xylem and phloem tubes.

Transpiration

Transpiration is the evaporation of water from the leaves of a plant. Before beginning to teach transpiration, ensure pupils are familiar with changes of state: remind pupils of water evaporating from a kettle. You can demonstrate transpiration by putting a plant in a plastic bag for a day before the lesson. Condensation (from the leaves) will form on the inside of the bag. If you forget to set it up, collect some condensation from a boiling kettle, and then put the bag around the plant. Conclude that the water must have evaporated from the plant, and that this evaporation is known as transpiration.

To prove the water came from the plant and not the soil, you can set up the apparatus in Figure 18.1 (use a balance accurate to three decimal places). During the lesson the mass of the plant will drop, because of water leaving the leaves.

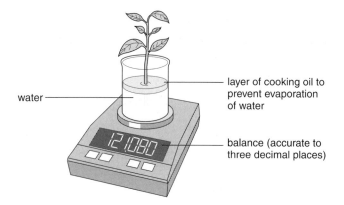

water — layer of cooking oil to prevent evaporation of water

balance (accurate to three decimal places)

Figure 18.1 *Apparatus to show that plants lose water from their leaves. The presence of the oil prevents evaporation from the water in the beaker, confirming any loss in mass must come from the leaves*

More water is lost from the lower surface of the leaf than the upper surface. Demonstrate this using two methods:

- Using tweezers, stick a piece of cobalt chloride paper to the top and the bottom of a leaf. The paper on the under side will turn pink, indicating the presence of water, more quickly than the paper on the upper side.
- Detach four leaves from a plant. Leave one untreated, put vaseline onto (i) both top and bottom surfaces, (ii) only the top surface, and (iii) only the bottom surface of the other leaves. The plants without vaseline on their bottom surface will lose water and wilt most quickly; this may take several days.

Pupils should relate the difference in water loss to the leaf's adaptations to prevent such loss. The waxy cuticle on the upper side of the leaf prevents loss of water. Pupils should know from examining leaf cross-sections that the holes in the underside are called stomata (singular: stoma). Water can evaporate from the cells inside the leaf into the air spaces, and then out of the stomata. Show that the majority of stomata are found on the underside of the leaf by making a cast using acetate and acetone. This procedure is difficult and pupils may need a lot of help.

- Take a leaf from any plant whose leaves are not hairy: privet (*Ligustrum vulgare*) works well.
- Dip it into acetone using tweezers, push its lower surface against the acetate with a rubber bung for at least one minute. Then carefully peel off the leaf.
- Look at the acetate under medium power: you will see the outline of lots of

stomata (they look like two bent sausages with holes in between). Try it with the upper side of the leaf and you will find fewer, if any, stomata.

Pupils should conclude that stomata are located on the underside of the leaf and are the route by which water leaves the plant by evaporation. Ask them why the stomata are situated on the underside of the leaf: this receives less incident sunlight than the top side, and water will not evaporate as quickly, hence conserving water.

If water is lost through the stomata, why have them? Pupils should realise that during the day, gaseous exchange needs to occur through the stomata because plants need carbon dioxide for photosynthesis.

Because photosynthesis does not occur at night, plants can reduce water loss by shutting their stomata. They can also shut the stomata when they are short of water. The mechanism by which this happens is based on osmosis. Only introduce it to more able pupils, who will still find it difficult to understand.

- The stomata are surrounded by guard cells. These are the only cells in the lower epidermis which contain chloroplasts and which photosynthesise.
- During the day, they photosynthesise, increasing the concentration of glucose in the cytoplasm.
- This decreases the relative concentration of water molecules in the cytoplasm of the guard cells.
- As a result, water flows into the cells from neighbouring lower epidermal cells.

Most pupils will predict that the guard cells will bulge out, and may therefore suggest that the stoma would close as a result; in fact, the stoma opens. Because the wall on the side of the stoma is thickened, it does not bulge out. The rest of the wall does bulge, and as a result the stoma is pushed open. Use a model to make this clear.

- Take two long, partially inflated balloons and stick a piece of Sellotape on one side of each balloon.
- Hold the balloons together so the Sellotaped sides are facing, and fully inflate both balloons.
- They will each form a semi-circle shape, hence leaving a hole between them (Figure 18.2).

Based on this mechanism, pupils should predict whether stomata will be open or closed when the plant is or is not photosynthesising. If you make a stomatal cast (see earlier) from a plant which has been kept in the dark for several days, and one which has been photosynthesising, you should see a difference in the imprint of stomata viewed under the microscope.

This mechanism also allows the plant to conserve water. Ask pupils what will

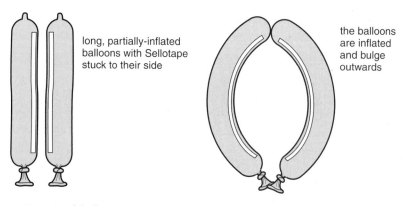

long, partially-inflated
balloons with Sellotape
stuck to their side

the balloons
are inflated
and bulge
outwards

Figure 18.2 *Model of a stoma*

happen to the guard cells if the plant does not have adequate water to diffuse
into them by osmosis: they will not bulge out, and the stoma will remain closed.
This prevents more evaporation of water from the leaf.

Assessing pupils' learning

- Pupils should write up experiments demonstrating water loss from the
  leaf and explain how their results demonstrate (i) that water is lost from
  leaves by evaporation, and (ii) that most water is lost from the underside
  of the leaf due to the prevalence of stomata.
- Higher level pupils should provide written understanding of why stomata
  need to be open during the day and closed at night. They should also be
  able to describe the mechanism by which the stomata are open or closed.

KEY STAGE 3 CONCEPTS

The transpiration stream

Pupils merely need to know that water flows from the roots, up the stem and out
of the leaves. They do not need to know how that flow is driven.

KEY STAGE 4 CONCEPTS

The transpiration stream

This is perhaps the most difficult section in this chapter. Explain to foundation
level students that water passes up the stem from the roots, and into the roots
from the soil. You can demonstrate water moving up the stem using celery as
outlined above for xylem. Explain that water moves into the roots and up the

stem to replace the water which is lost through transpiration from the leaves. If there is not adequate water to move into the plant, then the plant's cells become flaccid and the plant wilts. Leave a geranium (*Pelargonium*) without water for a week to demonstrate.

Higher level pupils must understand that water moves into the roots in response to loss of water from the leaf. To explain this, follow the path of water, starting at the leaves. Provide students with a diagram of the leaf cross-section, including a cross-section of a vein. Explain that water evaporates from cells into the air spaces within the leaf. This lowers the relative concentration of water in the cells next to these air spaces, and water moves by osmosis from neighbouring cells. This in turn lowers the relative concentration of water in these cells, and water moves into them from neighbouring cells. This continues, all the way back to the xylem vessels in the leaf.

Remind pupils that the xylem vessels effectively form a continuous tube, from the xylem in the leaf, down the stem, to the xylem in the root. If you remove water from the top of the xylem, water will move up the xylem by diffusion from a region of high water concentration in the root.

If water is removed from the xylem in the root, then water will diffuse by osmosis from the root hairs (where there is a relatively high water concentration) into the root xylem (where the relative water concentration has reduced because water has moved up to the stem), and from the soil into the root hairs.

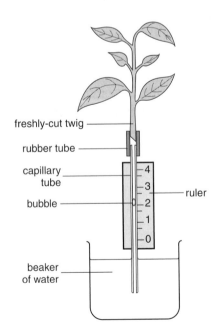

Figure 18.3 *A potometer to investigate the rate of transpiration. As water evaporates from the leaves, water moves up the tube and into the roots in response*

Demonstrate that evaporation of water from the leaves causes this transpiration stream by investigating the rate of water uptake using a simple potometer (Figure 18.3). In the winter, *Buddleia* is a good species because it is relatively frost-resistant. At other times of year, willow (*Salix*) and willow-herb (*Epilobium*) give good results. Use a sprig with lots of leaves and measure the rate of water uptake. Having done so, put vaseline on the lower surface of the leaves (the location of most stomata) and measure the rate of water uptake again. Because less water is being lost from the leaves, less water will be taken up by the roots.

Investigate the factors which affect the speed of transpiration by blowing hot air onto a plant with a hair dryer (investigating the effect of temperature and wind) and placing the plant in a plastic bag (investigating the effect of a humid atmosphere). More able pupils should explain these results according to their effect on the rate of diffusion of water out of the leaf.

Assessing pupils' learning

Pupils should

- explain how diffusion of water vapour out of the leaf causes movement of water up from the roots. This process is often found very difficult to grasp and you could give pupils statements or pictures to place in the correct order.
- write up the potometer experiments, describing how the leaf is involved in moving water up the plant.

References

Burton, I.J. (1999) A simple technique for preparing liver cells for microscopical examination and a description of their structural features. *Journal of Biological Education* **33**:113–114

4 Variation, Inheritance, Evolution and Classification

chapter 19

Variation

BACKGROUND

All living things are different. Members of different species look different, live in different ways and cannot interbreed. Some species are more similar than others: for instance, a human and a lion have less in common than a human and a monkey. This reflects the way in which they evolved (see Chapter 21) and is reflected in the way that species are classified (see Chapter 22).

There is also variation in characteristics within a species. This may be because different individuals inherit different characteristics from their parents (genetic variation) or because the environment in which particular individuals live affects their characteristics (environmental variation). The factors which control inherited characteristics are genes (the instructions which run the cell and determine the characteristics of the body) which are found in the nucleus arranged in string-like structures called chromosomes. Each person has two copies of each gene (e.g. eye colour gene) which may be the same (e.g. two blue genes) or different (e.g. a brown and a blue gene) (see Chapter 3). When gametes are formed, one copy of each gene goes into each sperm, or into each egg. When an egg and sperm from two different people fuse, this makes a cell with a new combination of genes. This new combination produces a new unique individual. The combination of genes in the gametes can be altered because of (i) random segregation of chromosomes at meiosis, and (ii) mutation.

Pupils arriving from KS2 will appreciate that species are different to one another, and because of these differences can be split up into groups. At KS3, you will need to help pupils to appreciate the differences between individuals of the same species. At KS4 double and single award, pupils must also appreciate the genetic and environmental causes of variation, that asexual reproduction makes identical offspring, that sexual reproduction causes variation, and that mutation (changes in genes) is a cause of variation. Most pupils are happy to accept the environment as a cause of variation, but find it more difficult to understand that sexual reproduction causes variation. Ensure you find out whether students have studied meiosis before you begin this topic, and assess their understanding of genes and the process of cell division.

KEY STAGE 3 CONCEPTS

Variation between species

Begin by asking pupils to define a species. Most pupils will identify morphological similarity as being important in distinguishing between species. Although this is adequate, try to challenge it. Give pupils two pairs of species which look very similar: e.g. human and monkey, zebra and horse. Suggest that although morphological similarity is important in defining a species, it is difficult to draw a line to say where one species begins and another ends. Tell pupils that individuals are only members of the same species if they can breed successfully with each other. Sharp pupils may ask about horses and donkeys (which are different species) but which can interbreed to produce mules. This is true, but the mules themselves are sterile: this does not therefore count as successful reproduction. Finish by concluding that, although the ability to interbreed is the 'definition' of a species, in practice, scientists often find it easier to distinguish between species according to morphological characteristics.

Assessing pupils' learning

- Give pupils a selection of plants and ask them to distinguish between species. Deliberately include different-sized examples of the same species, and examples of the same species which are in flower and not in flower.
- Ask pupils to define a species, and distinguish between pairs of animal species.

KEY STAGE 4 CONCEPTS

Variation between species is not included at KS4. However, much information from KS3 is required as a basis for classification at KS4 (Chapter 22).

KEY STAGE 3 CONCEPTS

Variation within a species

Characteristics within a species can vary considerably. A good example of variation is fingerprints: get pupils to dip their fingers into talcum powder and then place them onto Sellotape. Stick the Sellotape to black card. Discuss with pupils how you could measure the variation in fingerprints: conclude that it is difficult.

Move on to a variable which is easier to measure. Ask pupils to measure their height and collate the class results on the board. They can then produce a histogram of height. They could process the data and draw the histogram using a spreadsheet. For a less able group, use shoe size as a variable; they will not waste time measuring and the histogram will be easier to draw. Conclude that variation within a species normally exists between top and bottom limits. This is the so-called *normal distribution* of variation for a characteristic in a species. A different species may have a different normal distribution of variation for any characteristic.

> ### Assessing pupils' learning
>
> Ask pupils to compare the same characteristic between different species. They could suggest reasons why such characteristics differ between species with different ways of life.

KEY STAGE 4 CONCEPTS

Continuous and discontinuous variation

Pupils should produce a histogram of a continuous variable (e.g. height). Tell pupils that any variation which can be measured on a sliding scale, such as height, finger length etc., is known as continuous variation. Such variation always forms a normal distribution. Also ask them to produce a bar graph of a discontinuous variable: e.g. eye colour, ability to roll the tongue, blood group. Tell them that any type of variation which falls into groups (with no intermediates) is discontinuous variation. There are two nice examples of discontinuous variation which inspire pupils' interest: colour blindness (your department may have colour blindness testing cards), and the ability to taste phenylthiocarbamide (PTC).

Safety Advice: PTC in solution, if taken in excess, is toxic. Philip Harris produces strips impregnated with PTC which are generally considered safe. Pupils should not share tasting strips.

Sharp pupils may challenge you about eye colour being discontinuous as some people appear to have eyes which are a mix of colours. If they are referring to hazel eyes, tell them that hazel is actually a distinct colour and that there can be no intermediates between brown, hazel and green. If they are referring to patches in the iris, just tell them that this is a genetic anomaly. Although not entirely true, it will keep the story simple.

Assessing pupils' learning

- Ask pupils to distinguish between examples of continuously and discontinuously varying characteristics.
- Ask pupils to identify which characteristics on a plant are continuously or discontinuously varying.
- Less able pupils could produce a human body poster, identifying as many different ways in which that body could vary as they can, and labelling each type of variation as continuous or discontinuous.

KEY STAGE 3 CONCEPTS

Variation within a species: genetic and environmental

Remind pupils that they look like their parents. Tell pupils that genes control what a person looks like and that characteristics are inherited from your parents through genes: hence, genetic variation. If pupils ask why they have a different eye colour from either their mum or dad, you can explain about dominant and recessive genes (see Chapter 20). Do so only with the most able pupils.

Environmental variation is more difficult for pupils to understand. If you have identical twins in your school, they provide an excellent discussion point. They have exactly the same genes: they inherited exactly the same characteristics from their parents. However, if one eats lots and the other does not, they will differ in weight and height. If one played rugby yesterday they may have a scratch on their knee. If one spends more time out in the Sun, they may be more tanned. Ask pupils to think of as many sources of such environmental variation as possible.

Demonstrate the effects of genetic and environmental variation using plants. Get pupils to grow rapid-cycling brassicas from seed under standard environmental conditions. Because they have been subject to exactly the same environment, any variation between them must be due to genetic variation. Pupils could quantify this variation (e.g. number of leaves, internodal length, leaf size, etc.). To demonstrate environmental variation, take cuttings from one plant and give them two weeks to root properly. Most pupils will realise that these cuttings are genetically identical because they came from the same parent plant. You can then change the environment for each: e.g. put one in the dark, give one fertiliser, put one in a greenhouse etc. Ask pupils to quantify the differences caused by these environmental changes.

> ### Assessing pupils' learning
>
> - Give pupils a variety of types of variation, and ask them to identify whether they are caused by genetics, the environment or both. For example, height may depend on genes and diet.
> - Pupils should write up the plant practicals, showing they understand how variation was caused.

KEY STAGE 4 CONCEPTS

Genetic and environmental variation

Remind pupils of the distinction between environmental and genetic variation, using those activities described above. The mechanism by which genetic variation is caused is split between sections of the National Curriculum. Formation of gametes, which transfer genes to the offspring, is outlined in Chapter 3. If you have a choice, discuss formation of gametes first, and then how such formation can cause genetic variation. Bear in mind that because of the complexity of much of this material, it is greatly simplified, or omitted from most GCSE syllabuses, for foundation level students.

Remind pupils that each cell in a human contains 23 pairs of chromosomes. More able pupils should remember that cells in the testes and ovaries (gonads) divide by meiosis, such that only one of each pair of chromosomes goes into each sperm or egg, respectively. Remind pupils that the sperm and egg fuse in fertilisation and that the resultant zygote divides repeatedly to form a new embryo and foetus.

The causes of genetic variation in offspring are three-fold. Pupils invariably find these difficult.

1 Random segregation of chromosomes

Foundation pupils should know that the chromosomes in each pair are slightly different. They may carry different versions (alleles) of the same genes. When a sperm or an egg is formed, one chromosome of each pair is taken at random to be included in that sperm or egg. When an egg fuses with a sperm from someone else, the zygote formed contains a unique combination of genes.

For higher level pupils, begin at meiosis and show the chromosomes lined up in pairs across the middle of the cell. Explain that because each pair of chromosomes can line up randomly, the paternal origin chromosomes (the chromosome of each pair originally inherited from the father) could be transferred to a gamete together (Figure 19.1a), or they could be split up (Figure 19.1b). Because there are 23 pairs of chromosomes, and each pair can line up in two orientations, this is a source of genetic variation in gametes and therefore in offspring.

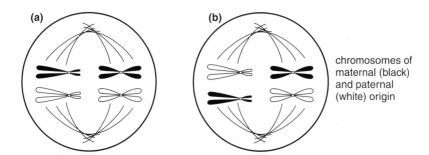

chromosomes of maternal (black) and paternal (white) origin

Figure 19.1 *Segregation of chromosomes during meiosis*

2 Random fertilisation of offspring

Because each gamete contains different genes, and any of the gametes could fuse with each other, this introduces another source of variation in offspring.

3 Mutation

Return again to meiosis, but to interphase when the chromosomes are being copied. If this copying goes wrong, a gene may be altered: this usually happens when the order of bases in the genetic code is changed. Use some examples of words which are spelt similarly to get the message across. For instance, just one mistake can change 'look' to 'book'. This changes the meaning of the word, and likewise, a change in the genetic code causes the effect of the gene to change. For instance, a mutation in an eye colour gene may cause the colour to change from brown to blue.

Stress that this is a natural process, but it can be made more likely by some chemicals and ionising radiation which are collectively called mutagens. Mutagens cause cancer. You could give more able pupils a research project to find out as many different causes of mutation/cancer as possible.

Assessing pupils' learning

- Ask pupils to write in the first person as if they are a gene on a chromosome. They should outline what happens to them as the cell divides, with which other chromosomes they pass into a sperm, and what happens to them upon fertilisation.
- Pupils could produce a wall display outlining the causes of genetic variation.

chapter 20

Inheritance

BACKGROUND

Every cell in the human body has two versions of each chromosome. Because chromosomes are strings of genes, each cell also has two copies of each gene (alleles). These may be identical versions, or they may be different. For any particular gene, and therefore any particular characteristic, (i) if both alleles in each cell are the same, that person is referred to as being homozygous for that characteristic; (ii) if they are different, that person is referred to as being heterozygous. When an egg and sperm fuse in fertilisation, the new person gains one of each pair of chromosomes, and therefore one allele for each characteristic from each parent.

If a new baby has two copies of the same allele (e.g. brown eye colour), they will have brown eyes. If they have two different alleles (e.g. blue and brown), they will still have brown eyes. This is because the brown allele is dominant to the blue allele; the blue allele is recessive. The mechanism underlying this process is simple. Blue is the default eye colour and blue eyes occur unless the brown allele is present. The alleles for a particular gene possessed by an individual are referred to as the genotype. The effect of the genotype on the body's outward characteristics is called the phenotype.

Gender of offspring is determined by the 23rd pair of chromosomes: the X- and Y-chromosomes. Males have one X- and one Y-chromosome in every cell, and can produce X-sperm or Y-sperm. Females have two X-chromosomes in every cell and can only produce X-eggs. The sex of an offspring is determined by whether an X- or Y-sperm fuses with the egg. Some inherited diseases are caused by genes on the normal chromosomes, and some by genes on the sex chromosomes. Selective breeding can be used to produce individuals with desired characteristics, utilising the fact that phenotypic characteristics are passed on through inheritance of genes. Genetic engineering achieves the same ends, and cloning allows us to utilise the products of specific genes on an industrial scale.

Pupils will arrive from KS2 knowing simply that they look like their parents. Some may understand that they have inherited their characteristics. At KS3, pupils must understand that characteristics can be passed from one generation to the next, and that this can be utilised in selective breeding programmes. Misconceptions are rife in this area, including: (i) male offspring inherit characteristics from their dad, and female offspring from their mum, and (ii) one parent is dominant in the transmission of characteristics to offspring. Remember that pupils know about genes, chromosomes and meiosis from Chapters 3 and 19. Pupils at KS4 double and single award must understand (i) how to carry out a monohybrid genetic cross, (ii) how sex is inherited, (iii) that some diseases can be inherited, and (iv) the principles of selective breeding, genetic engineering and cloning. Research has shown that pupils have a broad general knowledge about genetic engineering, and some students have significant ethical opinions. Because of this, it is a subject which can provoke much interest. Be aware that in many syllabuses, the depth with which all these subjects is considered is sparse for foundation level students.

KEY STAGE 4 CONCEPTS

Monohybrid inheritance

Monohybrid inheritance looks at the inheritance of one gene on a particular chromosome. Remind pupils that:

- chromosomes are strings of genes and that chromosomes occur in pairs in each cell.
- there are therefore two of each gene in each cell.
- these may be identical (e.g. both blue eye colour genes) or may be different (e.g. brown and blue) versions (alleles) of the gene.

At this point, introduce pupils to the definitions of homozygous (two identical alleles for a particular characteristic in each cell), heterozygous (two different alleles for a particular characteristic in each cell), genotype (the alleles which a person has in their cells for a particular characteristic) and phenotype (the outward effect of those alleles). Having done so, introduce the concepts of dominance and recessiveness. In the case of eye colour, brown overpowers green; anyone with a green allele and brown allele has brown eyes. Try making this clear with the following analogy (Dolan 1996):

- Imagine a gene for hat type with two alleles: baseball cap and witches hat.

- If a pupil only wears a baseball cap, their phenotype is quite clearly 'baseball cap'.
- If the pupil puts on the witches hat as well, on top of the baseball cap, it masks the baseball cap. The baseball cap is still there, but we can't see it anymore, and the phenotype has changed to witches hat.

Gregor Mendel, an Austrian monk, originally introduced the ideas of dominance and recessiveness. Your syllabus may require knowledge of Mendel's work which is outlined in almost every biology textbook.

Introduce your first genetic cross (Figure 20.1), working through it step-by-step. Pupils will reinforce their understanding if they include all components of the cross, literally word-for-word. If they rote-learn the wording of the cross, they will also gain full marks at GCSE! Ensure that when they write the gamete genotypes, they separate the letters (each allele being in different gametes). When defining the code letters for genes, ensure that the dominant has a capital; convention says this letter must be the first letter of the dominant allele (e.g. brown $= B$). The letter representing the recessive allele should be the corresponding small letter (e.g. green $= b$). Note that, by convention, the recessive allele in this case is not g for green.

A homozygous woman with green eyes marries a heterozygous man with brown eyes. Brown eyes are dominant to green eyes. Predict the genotypes and phenotypes of their children using a genetic diagram.

Let B = gene for brown eyes

Let b = gene for green eyes

Brown eyes are dominant to green eyes.

Parents' phenotypes: Green eyes × Brown eyes

Parents' genotypes: bb × Bb

Gametes' genotypes: b b B b

Fertilisation and offspring genotypes

	B	b
b	Bb	bb
b	Bb	bb

Offspring phenotypes: 50% brown eyes

                          50% green eyes

Figure 20.1 *Genetic diagram of a typical monohybrid genetic cross*

Make clear that the parent genotype refers to:

- the alleles for a particular characteristic
- present on a pair of chromosomes
- in each cell in the parent's body.

When one of these cells in the gonads divides by meiosis, it produces gametes, each containing one of these alleles. Because fertilisation occurs at random (see Chapter 19), either of the two sperm produced in the testis could fertilise either of the two eggs produced in the ovaries.

You may find it useful to use poppit beads to model the pair of chromosomes upon which your alleles are located, and to go through a cross with these chromosomes on a central bench. To reinforce the idea of random fertilisation, use different coloured beads to represent different alleles, and place the correct proportion of male gamete beads into one beaker, and female gamete beads into another beaker. With eyes closed, pupils can choose one of each to simulate fertilisation. If pupils do enough random fertilisations, the genotypes produced should be in the ratio predicted by the cross.

Pupils sometimes find it difficult to interpret the results from crosses. Tell them there will always be four genotypes produced by a cross. If four different genotypes are produced by a cross, then the couple have a one in four chance of having a child with either of these genotypes. They must translate the genotypes back to phenotypes using what they know about the dominant or recessive alleles. If two of the possible offspring genotypes are the same ($bB = Bb$), then there will be a 50% chance of getting offspring with that genotype.

Some pupils think that parents having four offspring will definitely have all four genotypes of offspring. Stress that for each child there is a one in four chance of having an offspring with a particular genotype. A couple's first four offspring *could* all have the same genotype!

It is difficult to include practical work in this topic because it takes some time for most animals and plants to reproduce, and for offspring to show their phenotype. Price & Harding (1993) suggest using rapid-cycling brassicas to investigate monohybrid inheritance, and Shaw (1997) has described a game to assist learning.

Assessing pupils' learning

- Pupils must demonstrate they understand the meaning of homozygous, heterozygous, genotype, phenotype, dominant and recessive. Ask pupils to match up cards with the names to the definitions.
- Pupils only really understand monohybrid inheritance when they have carried out some of their own genetic crosses. Your department will probably have some question sheets. There are two types of question: one

which tells pupils which alleles are dominant and recessive, and one where they must work it out from information provided.

- Check pupils understand how genetic crosses relate to the alleles within the cell. Ask them to write in the first person as if they are one of the alleles. They should explain their location in the cell, and what happens to them at meiosis and fertilisation.

Inheritance of sex

Begin again at the cell and the 23 pairs of chromosomes contained within the nucleus. The 23rd pair of chromosomes are the sex chromosomes. Tell pupils these chromosomes are called X and Y. Males have an X- and a Y-chromosome in each cell. Females have two X-chromosomes. This is likely to be enough detail for foundation level pupils.

More able pupils should be able to write out a genetic cross to show the inheritance of sex (Figure 20.2). Bear in mind that neither X nor Y is dominant (they are chromosomes, not genes). The presence of both an X- and a Y-chromosome in a person's cells makes them male. The presence of two X-chromosomes makes them female. Again, pupils should be able to follow the genetic cross and write out each stage. Use questions to make pupils realise that the sperm determines the sex of the offspring: it is only in the sperm that the sex chromosomes differ.

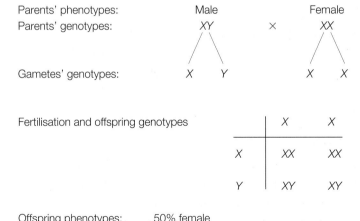

Figure 20.2 *Genetic diagram showing the inheritance of sex*

Inherited diseases

There are two kinds of genetic disease. Some are caused by recessive alleles on normal chromosomes, e.g. sickle-cell anaemia and cystic fibrosis. Explain their inheritance using simple monohybrid genetic crosses. When someone has heterozygous parents and gains two recessive alleles, they contract the disease. Heterozygotes (i.e. people who have one recessive allele) are referred to as carriers, because they have the allele for the disease but do not develop it because the normal allele is dominant to the disease allele.

The second type of genetic diseases are those caused by recessive genes on the X-chromosome. Tell pupils that the X-chromosome is longer than the Y. This means that for genes located at the end of the X, there is no equivalent gene on the Y (Figure 20.3).

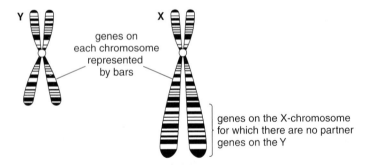

genes on
each chromosome
represented
by bars

genes on the X-chromosome
for which there are no partner
genes on the Y

Figure 20.3 *The X- and Y- chromosomes. Some genes on the X-chromosome have no equivalent gene on the Y-chromosome because the Y-chromosome is shorter*

Hence, if a gene on the X-chromosome of a male is faulty, there is no normal dominant gene to mask its effect and a disease can develop. Genetic disease can also be contracted by females, but this is rarer because the other X-chromosome can have a dominant normal copy of the gene. Colour-blindness, Duchenne muscular dystrophy and haemophilia are caused by recessive genes on the X-chromosome, 'so-called' X-linked recessives. Some syllabuses require knowledge of the symptoms of these diseases which you can find in most textbooks. In addition, some syllabuses require higher level pupils to show the products of genetic crosses involving X-linked recessives. Figure 20.4 shows a cross using haemophilia as an example.

Let N = normal gene
  n = gene for haemophilia

N is dominant to n

Parents' phenotypes: Normal female   ×   Normal male

Parents' genotypes: X_NX_n × X_NY

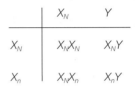

Gametes' genotypes: X_N X_n X_N Y

Fertilisation and offspring genotypes

	X_N	Y
X_N	X_NX_N	X_NY
X_n	X_NX_n	X_nY

Offspring phenotypes: 50% normal female
                          25% normal male
                          25% haemophiliac male

Figure 20.4 *Genetic diagram showing the inheritance of haemophilia, a sex-linked disease*

Assessing pupils' learning

Ask pupils to
- complete genetic crosses for genetic diseases caused by recessive alleles carried on the autosomes (non sex chromosomes).
- complete genetic crosses for genetic diseases carried on the X-chromosome.
- explain, using genetic crosses, why males are more likely to develop X-linked genetic diseases. To contract haemophilia, females would have to inherit an X-chromosome which carries the recessive haemophilia allele from both parents. For males to contract haemophilia, they just need to inherit an X-chromosome with the allele from their mother.

KEY STAGE 3 CONCEPTS

Selective breeding and genetic engineering

Begin by putting pupils in the position of a farmer. He has one cow which produces vastly more milk than the others. How does he get more cows which are good at milk production? Most pupils will suggest breeding using that cow. Put pupils in the position of a flower grower. He wants to combine tallness with yellow flowers. How does he combine both in the same plant? Breed the two together.

You will have to tell pupils that the sought-after characteristics may not appear in the first generation. You may have to breed this generation again before individuals of the correct phenotype result. You would then need to breed these individuals together to gain a population with the sought after characteristics. You can do selective breeding with populations of rapid-cycling brassicas (Collins & Price 1996). They vary in the distribution of hairs on the leaf surface, leaf margins, petioles and stems. Because plants flower within 14 days of being sown, you may have time to run selective breeding programmes over a whole term. Rapid-cycling brassica kits are available from Philip Harris Ltd.

Pupils should be aware of the range of uses for selective breeding: increasing arable and pastoral farming yields and producing pretty flowers!

Assessing pupils' learning

- Give pupils diagrams of plants with a variety of characteristics. Ask them to explain how to use selective breeding to create a population with designated characteristics.
- Ask pupils to make a board game: the aim could be to pick up desired characteristics by breeding.
- Ask pupils to think of as many agricultural uses of selective breeding as possible.

KEY STAGE 4 CONCEPTS

Selective breeding and genetic engineering

Recap pupils' understanding from KS3. They should realise that any characteristic being bred for must be controlled by an allele, either dominant or recessive. By selectively breeding, these alleles are perpetuated into the next generation.

Ask pupils to consider the advantages and disadvantages of selective breeding.

Possible disadvantages include: (i) takes a long time, (ii) for recessive genes it is difficult to follow the path of the gene because it is masked by the dominant allele, (iii) careful records of breeding must be kept, (iv) lots of individuals are produced which are genetically similar – if a disease can affect one, it can affect them all, (v) it is difficult to tell whether variation is genetic or environmental.

The aim of selective breeding is to get particular genes together in the same individual. Genetic engineering provides a quicker way of doing this. A gene can be inserted into a bacterium's DNA. The bacterium then produces the gene's products. If you multiply the bacteria, they can be used industrially. Insulin is produced in this way for treatment of diabetics. It is possible to inject bacteria carrying the desired gene into an individual and for the bacteria to insert their DNA, including the gene we want, into their hosts cells. Having understood the method, pupils could consider what may go wrong (e.g. the wrong gene may be transferred; a gene may end up in the wrong plant: a gene for high yield may be passed from a genetically-engineered crop into weeds by cross-pollination).

Assessing pupils' learning

- Ask pupils to write out the genetic crosses they would carry out to produce a population of individuals with a particular characteristic, given a range of potential parents. Ask them to breed for characteristics which are controlled by a dominant gene, and a recessive gene.
- Pupils should outline the method involved in genetic engineering as a flow chart; they could write in the first person as if they were a gene describing what happened to them during genetic engineering.
- Pupils should write an editorial, arguing whether the benefits of genetic engineering outweigh the risks.
- Set up a debate about the ethics of genetic engineering. Lucassen (1995) describes a useful format for organising such a debate.
- Ask pupils to find any newspaper articles on genetic engineering and to review them to see if they are based on good science.
- Pupils could produce an information leaflet to inform the general public about genetic engineering.

Cloning

Cloning is the production of a genetically-identical individual. Many living things do it themselves: by asexual reproduction (see Chapters 10 and 16). Use a spider plant to remind pupils of this: it produces runners upon the end of which new plants grow. The runner withers away and the new plant takes root, producing a new plant genetically identical to its parent: a clone.

Ask pupils to think of the advantages of cloning: relatively quick (it takes a

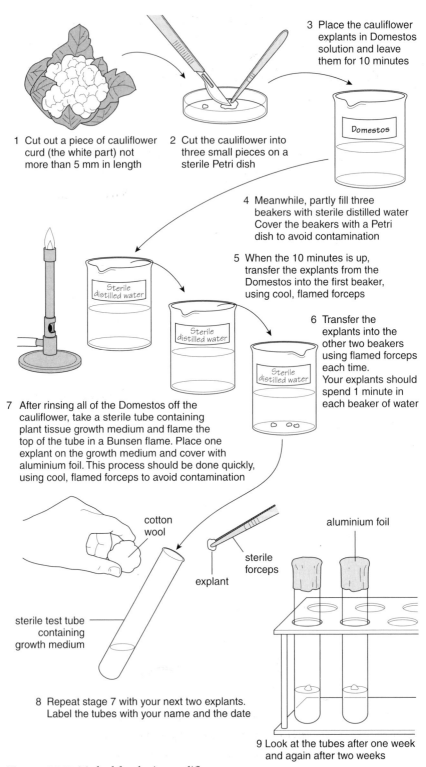

1 Cut out a piece of cauliflower curd (the white part) not more than 5 mm in length

2 Cut the cauliflower into three small pieces on a sterile Petri dish

3 Place the cauliflower explants in Domestos solution and leave them for 10 minutes

Domestos

4 Meanwhile, partly fill three beakers with sterile distilled water Cover the beakers with a Petri dish to avoid contamination

5 When the 10 minutes is up, transfer the explants from the Domestos into the first beaker, using cool, flamed forceps

Sterile distilled water

Sterile distilled water

Sterile distilled water

6 Transfer the explants into the other two beakers using flamed forceps each time. Your explants should spend 1 minute in each beaker of water

7 After rinsing all of the Domestos off the cauliflower, take a sterile tube containing plant tissue growth medium and flame the top of the tube in a Bunsen flame. Place one explant on the growth medium and cover with aluminium foil. This process should be done quickly, using cool, flamed forceps to avoid contamination

cotton wool

aluminium foil

sterile forceps

explant

sterile test tube containing growth medium

8 Repeat stage 7 with your next two explants. Label the tubes with your name and the date

9 Look at the tubes after one week and again after two weeks

Figure 20.5 *Method for cloning cauliflower*

whole season to produce a flower and then a fruit and for that fruit to be dispersed and grow) and only one parent is required (a plant does not need to wait to be fertilised). There is a disadvantage: that all individuals will become genetically identical and may all be susceptible to the same diseases.

Cloning is used commercially. Ask pupils to imagine they have engineered (either by selective breeding or genetic engineering) their ideal individual. If they then have to breed this individual again, they will lose some of that perfection. To be able to clone that individual would avoid such loss of perfection. There are two ways: taking cuttings and tissue culture. You can get less able pupils to take some cuttings from geraniums. Higher ability pupils could try tissue culture. A method to clone cauliflower is described in Figure 20.5.

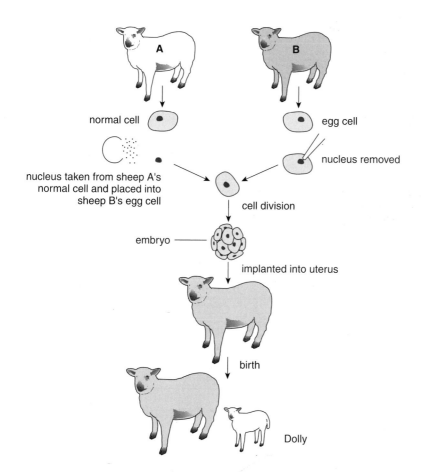

Figure 20.6 *Method for cloning animals. An egg cell is taken from Sheep B and its nucleus removed. This is replaced with the nucleus of a cell from Sheep A. The resultant cell divides to form an embryo and is implanted back into Sheep B's uterus from where it will develop into a new lamb: a clone of Sheep A*

Within the last few years, Dolly the sheep has been produced by cloning. Although this is unlikely to be included on a GCSE syllabus, the method is outlined in Figure 20.6 if you have a particularly interested group. Clones *can* occur in humans; identical twins are the result if an embryo splits in half soon after fertilisation.

Assessing pupils' learning

- Pupils should write about how a spider plant reproduces. More able pupils must decide whether mitosis or meiosis is involved in the production of the new plant, and whether there is any opportunity for variation to be produced.
- Write a newspaper article about the advantages and disadvantages of cloning for commercial products. Pupils could think about the implications of the technology for cloning humans.
- Pupils should write up cuttings and tissue culture experiments. They should demonstrate that they know the daughter plants are growing by mitosis: i.e. no variation produced.
- Ask pupils to search the press for articles on cloning. They should review them to establish the accuracy of the science.

References

Collins, N. & Price, R. (1996) *Plants.* In: Living biology in schools, M. Reiss (ed.). Institute of Biology, London

Dolan, A. (1996) The making of Mandy – introducing alleles. *Journal of Biological Education* 30:94–96

Lucassen, E. (1995) Teaching the ethics of genetic engineering. *Journal of Biological Education* 29:129–138

Price, R. & Harding, S. (1993) Genetics in the classroom: inheritance patterns of two mutant phenotypes in rapid-cycling *Brassica rapa. Journal of Biological Education* 27:161–164

Shaw, A. (1997) Alien alleles. *School Science Review* 78:284

chapter 21

Evolution and natural selection

BACKGROUND

Charles Darwin realised, from looking at the fossil record, that species changed gradually over time – they evolved. He also realised that species are adapted to their environment. At the time (1859), people still believed that God created all species. Darwin could not understand, in that case, why God had allowed some species to die out, as was apparent from the fossil record. He suggested that the variation between members of the same species allowed some to survive slightly better than others in their environment. Those least adapted to their environment do not reproduce and die. Those well adapted survive and reproduce. This is called 'survival of the fittest'. Because individuals' characteristics are controlled by genes, those genes which are beneficial to survival are passed on to the next generation, and this generation is also well adapted to its environment. Because nature (the environment) is selecting which organisms survive and pass on their genes, this process was called by Darwin 'evolution by natural selection'.

Pupils arriving from KS2 and 3 will have no prior knowledge of evolution from school. They may have heard about evolution from the television and you should be aware that some pupils will have some understanding. At KS4 double and single award, pupils must understand that evolution is a gradual change in a species, and that this is caused by selection acting on variation in a species. They should also understand how the fossil record provides evidence for evolution.

KEY STAGE 4 CONCEPTS

What are fossils?

Most pupils will know that fossils are old. Some will suggest they are bones or shells turned into stone. *Jurassic Park* has been responsible for broadening childrens' knowledge of fossil formation, and some pupils will have gleaned a fairly good understanding as a result. Your department may have some fossils to show pupils. Be aware that many fossils are likely to be fragile. You should discuss the methods of fossil formation:

- *Impression.* Ask pupils how sedimentary rocks are formed. Most should have encountered this in chemistry. If a dinosaur has left impressions (footprints) in mud and sedimentary rock forms over the top, the impressions will be preserved when layers of sediment squash the mud until it is compacted enough to be called a rock.
- *Casting.* Ask pupils to squash a leaf from above and below with plasticine, to simulate being engulfed by sediment. The leaf will eventually rot away and will leave a space in the rock with the surface impression of the leaf imprinted onto the surface of the cavity (a mould). When someone breaks open the rock, they see the imprint. Sometimes, water rich in mineral salts will flow into the cavity, and rock will be laid down within this cavity forming an exact rock copy of the original object (a cast). If pupils find it difficult to see how water can 'carry' rock, remind them of stalactites forming in a cave. Pupils can make plaster of Paris casts from plasticine moulds of shells or small bones.
- *Petrification.* Occasionally, water rich in mineral salts will get into a specimen before it begins to decay. The lattice-like structure of bones makes it particularly easy for mineral salts to be laid down within them.
- *Whole preservation.* Pupils will remember the insect preserved in amber (from *Jurassic Park*) and some will be familiar with frozen woolly mammoths in the Siberian ice.

Assessing pupils' learning

Ask pupils to describe the methods of fossil formation.

Evidence for evolution

In 1859, most people believed in creationism, i.e. God created every species in seven days, and each has been on the Earth ever since. Darwin, on his voyage to the Galapagos Islands in 1831, examined fossils in cliffs of sedimentary rock. Because the lower layers of the cliff were laid down first, older fossils would necessarily be trapped in layers at the base of the cliff. Based on these age differences, Darwin noticed the following evidence for evolution:

- there is a general increase in complexity moving from older to younger fossils
- species we see today are not present in the fossil record of millions of years ago
- some species no longer exist, suggesting creation does not explain the origin of species perfectly.

Some pupils may believe the creationist explanation. Because creation is based purely on faith, pupils should feel under no threat from evolution, an explanation based on evidence. Having said that, more able pupils should appreciate that the fossil record does not yet provide a complete picture of evolution because:

- only the hard parts of organisms are fossilised
- many fossils have been destroyed by erosion
- many fossils have not yet been discovered.

Assessing pupils' learning

- Set up a debate between the creationists and the evolutionists. There are flaws with the evidence in the fossil record, and the creationists could use these to attempt to defeat Darwin's idea of evolution. For example, there are many gaps in the fossil record which do not show the evolution of each species from millions of years ago through until now.
- Alternatively, ask pupils to write an editorial for *The Times* of 1859, describing the difference between creationism and evolution and arguing for one or the other.

Evolution by natural selection

How do species evolve? Many pupils will intuitively share the ideas proposed by Lamarck. He said that giraffes had stretched their necks over many generations to reach higher and higher leaves, passing on a stretched neck to their offspring. That is, characteristics which giraffes had acquired in their lifetime were passed onto their children. Explain why this is wrong – a baby does not inherit characteristics acquired in their parents' lifetime such as scratches or broken bones.

Darwin, on his journey to the Galapagos Islands, noticed that every species was perfectly adapted to the island on which it lived, and proposed a theory of natural selection. Imagine a leopard population:

- The fastest runner will catch more prey and survive to produce more cubs who will have the gene for fast running.

- Eventually, the whole population will contain the gene for fast running because all the slow runners will be out-competed for food.
- The population will have changed slightly: it will have evolved. Over many years, many such changes will accumulate until the population is so different from the original species that it will be labelled a new species.

With higher level pupils, generalise this process:

- There must be competition between individuals. If there was no competition, no individuals would die, and genes would not 'die out'.
- To make competition, there must be over-production of individuals e.g. leopards reproduce too quickly, producing a larger population than the environment can sustain.
- Individuals must vary; some are better adapted to their environment than others.
- Only the fittest survive. This does not mean those which are necessarily physically fit. Fit, in an evolutionary sense, means best-adapted. For instance, it may refer to fast running; it may also refer to giraffes having the longest necks to reach their food.

Assessing pupils' learning

- Give pupils a 'case-study' of evolution in action. Ask them to explain it using the words over-production, survival of the fittest, natural selection and variation.
- Ask pupils to explain why environmentally-caused variation is not important to Darwin's theory.
- Set up a debate between creationists and natural selectionists, or between natural selectionists and Lamarckists. Alternatively, in each case, ask pupils to consider the evidence for the two arguments in a newspaper article.
- Ask more able pupils to predict what may happen to European species if the European climate gets warmer gradually over a long period of time. Ask pupils to predict what may happen if the climate gets warmer over a very short period of time (species may not have time to adapt, and may become extinct).
- Ask pupils to explain the evolution of the peppered moth (*Biston betularia*). Some syllabuses require knowledge of the evolution of this moth, and this can be found in almost all biology textbooks.

chapter 22

Classification

BACKGROUND

Classification is used by biologists to place organisms into groups to make them easier to refer to. There are five kingdoms into which any organism can be classified: animals, plants, fungi, protists and prokaryotes. Each kingdom is divided into several phyla (singular: phylum) – or divisions in the plant kingdom – each phylum/division into several classes, each class into several orders, each order into several families, each family into several genera (singular: genus) and each genus into several species. The animal kingdom is divided into those organisms with or without a backbone: the vertebrate phylum and the invertebrate phylum, respectively. The vertebrates are divided into a number of classes including fish, amphibians, reptiles, birds and mammals. The plant kingdom is split into divisions. These include the green algae, the liverworts and mosses, the ferns and horsetails, the conifers and the flowering plants. When an organism is classified, whoever classifies it writes out a key (a series of questions) which will help any other biologist identify it.

Classification is not included at KS4. For KS3, teach classification after variation; pupils will already be familiar with the idea of a species. However, pupils arriving at KS3 will not be familiar with kingdoms, or dividing those kingdoms up into named groups. Pupils should be familiar with keys from KS2. These can cause less able pupils some difficulty, but if you approach them in a diagrammatic way, you should have no problem.

Research has revealed two common problems with learning about classification.

1 Pupils will accept the classification of organisms into particular kingdoms, phyla and classes etc. within lessons, but their own ideas about grouping living things (which they use in their everyday life) can be very difficult to overcome.

2 Pupils will classify vertebrates according to whether they have a definite outline and a definite head, rather than the presence of a backbone. Pupils often associate the concept of an invertebrate just with a crawling animal or one with a flat body, rather than with an animal which lacks a backbone.

KEY STAGE 3 CONCEPTS

The kingdoms

Begin by asking pupils to name or classify a mixture of plants and animals into two groups: plants and animals. Research has shown that some pupils have difficulty accepting trees and seeds as plants. They will accept that trees and seeds are part of the plant kingdom if you only give them two choices for classifying: i.e. plant or animal. Young or less able pupils may have difficulty identifying non-mammalian animals and humans as being animals. Again, tell them that all the species you have provided are either plant or animal. Most will then classify appropriately.

You can tell pupils the main difference between plants and animals: plants are autotrophic (make their own food using energy from the Sun) and animals are heterotrophic (must consume their food). Pupils should encounter at least a few examples of each of the other kingdoms. Set up a circus of pictures and specimens around the room, and ask them to classify them into the correct kingdom. The animals, plants and fungi will be no problem. Set up protists (amoebae are easy to get hold of: ask your technician) and prokaryotes (your department may have prepared microscope slides of bacteria) under the microscope and ask pupils to draw examples of each.

Having clarified the kingdoms, ask pupils to define some sub-groups of these based on the presence or absence of one characteristic (e.g. has petals, does not have petals; tail, no tail). They should keep dividing groups up until they only have one species in each group. They should then name each group and each species.

Next tell pupils about Carl Linnaeus. In 1735, he did exactly what they have just done. He gave each species two Latin names: a genus name and a species name – the 'so-called' binomial system for naming species.

- These two names should either be in italics when typed, or underlined when hand-written.
- The genus name always begins with a capital letter and the species name always has a small letter.

He also grouped genera into families, families into orders etc. These groupings were based on how similar groups of organisms were. For example, humans and monkeys are thought to be fairly similar and are placed into the order Primata. The primates are quite similar to the carnivores (lions and tigers) and are placed into the class Mammalia. This idea of nested hierarchies may cause some problems to younger or less able pupils. Use diagrams to show that humans lie within the primate group and all the primates lie within the mammals (Figure 22.1). Work through some examples to make this clear (see below).

	Humans	Lion	Sweetcorn
Kingdom	Animals	Animals	Plants
Phylum	Vertebrata	Vertebrata	Angiospermae
Class	Mammalia	Mammalia	Liliopsida
Order	Primata	Carnivora	Cyperales
Family	Hominidae	Felidae	Gramineae
Genus	*Homo*	*Panthera*	*Zea*
Species	*sapiens*	*leo*	*mays*

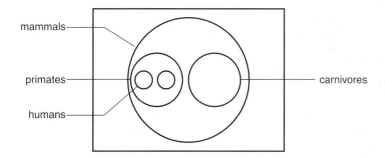

Figure 22.1 *A Venn diagram showing how groups of species are themselves grouped into larger groups*

Bear in mind that animals with amphibious habits (e.g. penguins and turtles) may be classified by pupils as amphibians. Animals which have English names like jellyfish and starfish may be classified by pupils as fish, whereas, because they lack a back-bone, they belong in the invertebrates.

Pupils often have difficulty remembering the names of the groups into which species should be classified. Try using the following mnemonic.

Kingdom	Kinky
Phylum	Pigs
Class	Can
Order	Only
Family	Fly
Genus	Going
Species	Sideways

Pupils should know the names of each class of vertebrates and the characteristics possessed by each. Provide pupils with pictures of members of each class (fish, amphibians, reptiles, birds and mammals) and ask them to identify the characters which define each group. Make them supplement conclusions with information from textbooks. If you can do fieldwork (see Chapter 23), or have a pond or tank with frogs and fish, ask pupils to identify the differences between live examples. More able pupils could research the orders into which one or more of these classes is divided. More able pupils could research the classes of plants.

Stress appropriate respect for animals and plants, and discourage pupils from picking wild flowers. You should not pick any flower without the land-owner's permission, and bear in mind that some species may be endangered.

S **Safety Advice:** If you do fieldwork, ensure you follow the safety advice outlined in Chapter 23.

Assessing pupils' learning

- Ask pupils to classify a set of nails and screws into hierarchical groups.
- Give pupils a variety of species and ask them to classify them into appropriate kingdoms. Include a variety of species from each kingdom.
- Ask pupils to classify species of vertebrate into the correct classes.

KEY STAGE 3 CONCEPTS

Writing and using keys

Linnaeus realised that he could name lots and lots of species, but that unless he wrote down which characteristics defined each species, no-one would be able to use his classification system.

- Ask pupils to think of two species each and write down a description of each on a piece of card.
- They should exchange pieces of card with other members of the class. The other member should try to identify which species they have described.
- Most pupils will be fairly unsuccessful in identifying other students' species.

Keys provide a better method for describing and identifying species. There are two types of key. Introduce pupils to diagrammatic dichotomous keys first (e.g. Figure 22.2).

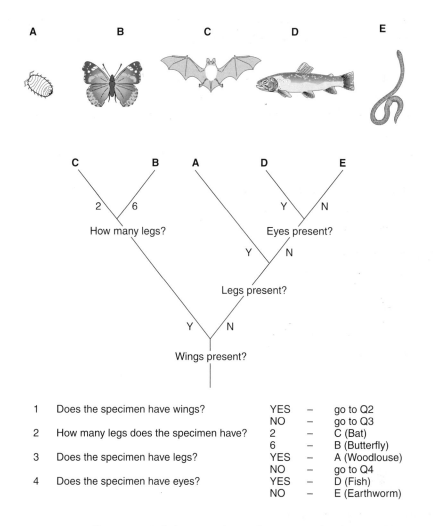

1	Does the specimen have wings?	YES	–	go to Q2
		NO	–	go to Q3
2	How many legs does the specimen have?	2	–	C (Bat)
		6	–	B (Butterfly)
3	Does the specimen have legs?	YES	–	A (Woodlouse)
		NO	–	go to Q4
4	Does the specimen have eyes?	YES	–	D (Fish)
		NO	–	E (Earthworm)

Figure 22.2 *A diagrammatic dichotomous key and a corresponding 'question' key*

- Give pupils six plant species and label them A to F. Pupils are unlikely to know the names of plants and will focus more on their morphological features. If given animals, less able pupils are often tempted to use a question such as 'Is it a robin?'
- At each node, pupils should ask a question for which the answer is either yes or no. The first node will split the plants into two groups, the next node will split these groups up etc. Eventually by this method you will differentiate between species.

The most common type of key used by ecologists, and which you should introduce to the more able pupils, is a question key. Instead of writing the options on a diagram, you simply take the alternatives from the diagram and write them out in question form (Figure 22.2).

Assessing pupils' learning

- Pupils should write a key to distinguish between a mixed selection of nails and screws. They should give each type of nail/screw a name, and write a key to help another reader work out which screw they have.
- Ask pupils to write a key distinguishing between playing cards. They will take great delight in being able to 'name that card' using only a series of questions. Set a challenge for the pupil who can identify any card correctly using the fewest questions.
- Set up a circus with different piles of objects spread around the room. Pupils should write a key to distinguish between objects in one of these groups. The keys should then be mixed up and given out again for another pupil to use.
- If you are doing fieldwork, give pupils examples from ponds, fields or gardens to distinguish between by writing a key.
- Your department may have keys written out for fieldwork activities which pupils can use for practice.
- Pupils could use a branching database program to create a diagrammatic key.

5 Living Things in their Environment

chapter 23

Adaptation, competition and predation

BACKGROUND

The environment in which a species lives determines its population size. The environment comprises both biotic (living) and abiotic (non-living) components. The abiotic include the rocks and soil, the climate and the availability of water. Biotic components include the availability of food, the presence of disease, competition for resources by individuals of the same or different species, and the level of predation. Species become adapted to the biotic and abiotic components of the environment by a process of natural selection (see Chapter 21). Adaptations are features which help species survive in their environment. Only those individuals best adapted to coping with these variables will survive. Such biotic and abiotic factors therefore have a strong influence on population size.

Pupils will know from KS2 that different animals and plants are found in different habitats. They will have studied the adaptations of animals and plants to two particular environments, although may not have met the word adaptation before. At KS3, pupils should understand that different animals and plants are adapted to living in different habitats, and that competition and predation will affect population size. At KS4 double and single award, they must understand how adaptation allows organisms to compete more effectively for resources, and how predator–prey relationships can limit population size.

KEY STAGE 3 CONCEPTS

The environment and adaptation

Most pupils will accept that population size depends on the number of organisms surviving, reproducing and dying in a particular area. Ask pupils to list the possible factors in the environment which may affect whether an individual survives or reproduces, and to explain what effect each factor may have on population size. Possible ideas may include temperature, sunlight, predators, pathogens/disease, and availability of water, oxygen and food. Summarise the activity by explaining that an individual organism's environment comprises living and non-living components.

To introduce adaptation, demonstrate that different organisms live in different habitats. If you are teaching this topic in the spring or summer, you can spend several lessons outside, catching and examining different species of animals and plants in different habitats. Use a spreadsheet to record the prevalence of species in different habitats. Draw bar graphs to see which species are most abundant in each habitat. Pupils should explain how adaptations of animals and plants are useful. For instance, a fish's gills allow it to obtain oxygen in water. Insects' jaws (mandibles) are useful for breaking up food.

Your department should have books describing field techniques for catching small animals. Do not allow pupils to pick any plants without the land-owner's permission and be careful what they do pick: some species may be endangered and it is an offence to uproot them. Ensure that you return any animals which you have caught back to their environment. Work hard to foster respect for the organisms being studied.

S Safety Advice: If you are going off-site, arrange appropriate insurance, parental permission, student–teacher ratios, etc. Your head of department or senior management team can advise. It is important to make yourself familiar with a fieldwork site before the pupils arrive. Ensure pupils wash hands after fieldwork and/or before eating. Although it is relatively rare in this country, ticks can carry Lyme's disease. If found burrowed into the skin, ticks should be removed by a doctor; do not try to pull them out yourself. There may also be a risk of contracting Weil's disease through cuts and grazes in contact with pond or river water. Ensure cuts are covered by plasters and check for prevalence of the disease with your local council.

If you cannot teach this topic in the field, bring in some examples of species which are highly adapted to their environments. Include some plants such as *Euphorbia*, a species which has many different varieties, all highly adapted to their environment. Botanical gardens often have plants which show very obvious

186

adaptations to their environment and are well worth a visit.

Having realised that species live in habitats to which they are suited, pupils must realise how some animals respond to daily and seasonal changes. Ask them when living things are most active. Most pupils will respond that most activity in living things occurs during the day. Species which are most active during the day are called diurnal species. Pupils may not appreciate that plants are more active during the day. Ask them when photosynthesis happens, and when plants bend towards the light. Pupils will realise that some species are nocturnal (more active at night) e.g. spiders, mice, bats and owls. More able pupils could consider the advantages of being nocturnal: e.g. it is less easy for predators to see them; they do not need to compete with diurnal species for food.

Seasonal changes in animals and plants are perhaps more obvious. Give pupils a circus of species showing seasonal adaptations: e.g. trees without leaves, a seed, birds migrating, a hedgehog hibernating, a root storing food (e.g. parsnip) and a rabbit with a thick coat of fur, and ask them to explain the adaptive value of these features. The RSPB produces some good resources about how birds cope with the winter.

Assessing pupils' learning

- Provide pupils with pictures of habitats (extreme habitats such as deserts and polar regions will produce the most obvious contrasts in adaptation), and ask them to list all the species which may live in that habitat, and explain how their adaptations help them to survive.
- Ask pupils to write a newspaper article about how one group of organisms is adapted to seasonal changes in their environment.
- Pupils could make a survival board game: for example, players could collect adaptation tokens for a particular habitat.

KEY STAGE 4 CONCEPTS

The environment and adaptation

Recap from KS3 the factors which make up a species' environment, list them under two headings, biotic and abiotic, and either ask pupils to define the words biotic and abiotic, or for less able pupils define them yourself.

Pupils should assess the relative abundance of different species in different habitats. Many syllabuses require knowledge of simple ecological field techniques. Your department will have a book outlining the majority of techniques. Quadrats and line transects can be used to describe changes in abundance of plant species according to the changes in the habitat. Traps (including pitfall traps) and pooters can be used to study different distributions of animal species in different habitats. Identification of many (especially insect) species is difficult. If you are not good at identification yourself, tell pupils to make up their own names for species, or simply call them species A, B, C etc.

When carrying out these techniques, more able pupils should understand why you must survey a habitat more than once to obtain a reliable estimate of population size of different species, and why sampling should be random. For instance, if the school field contained no daisies except in one corner, you would not provide a reliable estimate of population density by counting the number of daisies in quadrats thrown only in that corner.

Give pupils a range of different habitats to survey. They could use spreadsheets to show the relative distribution of different species in different habitats and should identify the adaptations that species have to their environments. They should also explain how these adaptations help survival. If you have a botanic garden near you, exotic plant species in the greenhouses show very obvious adaptations to their environments.

S **Safety Advice:** See the advice for KS3 above.

Assessing pupils' learning

- Ask pupils to describe the field techniques they have used and their limitations.
- Pupils should identify adaptations to the habitat in which each species lives and should try to explain the relative abundance of species in relation to their adaptations.
- Give pupils a variety of species and environments and ask them to place the species into the correct environments, based on their adaptations.

KEY STAGE 3 CONCEPTS

Competition

Competition between species and between individuals of the same species may affect population size. Recap that different habitats support different animals and plants because they have different adaptations. However, different species may use the same habitat in different ways (the way in which a species uses a habitat is called its niche). Species which live in the same habitat may therefore have different adaptations. Ask pupils to write down everything a human may do during a day and everything a cat does during a day. Make the list detailed, including exactly what they may eat, how they attract mates etc. The list should reveal that cats and humans, although sharing the same habitat, have different roles in that habitat: they have different niches.

At this point, conclude and define a niche. Ask pupils to imagine what would happen if the two species did share the same niche, or if their niches overlapped slightly. Would both species survive? More able pupils will realise that only those best adapted to their niche would survive and reproduce. Hence, the population size of this species would stay high. The size of the other species' population would reduce. For less able pupils use an example to help:

- A leopard and cheetah both hunt food by chasing after it.
- Individuals of the species which runs fastest will survive and reproduce.
- Individuals of the species which runs more slowly will fail to survive or reproduce.
- Because the cheetah run fastest, they will survive in that habitat; because leopards run more slowly, they will not survive.

Competition may also occur within a species. Not all members of a species are the same. For example, some leopards may run faster than others. If food supplies are scarce, only the fastest-running individuals will survive and reproduce.

Assessing pupils' learning

- Ask pupils to write about examples of competition in particular environments e.g. a garden.
- Choose a particular niche (e.g. Arctic, desert, tropical rain forest) and ask pupils to design a plant or animal perfectly adapted to that niche.

KEY STAGE 4 CONCEPTS

Competition

Recap the material from KS3. Pupils must be able to explain differences in relative abundance and distribution of organisms because of competition. Stress that organisms compete for mates, as well as resources. This will affect which, and how many, organisms go on to reproduce and will therefore influence population size and the variety of individuals in that population.

Assessing pupils' learning

Use similar activities to those used for KS3.

KEY STAGE 3 CONCEPTS

Predation

Predators are one of the biotic components of a species' environment. The ability of individuals to avoid predation affects whether they survive; predation therefore affects a species' population size.

- Give pupils a jumbled list of predators and prey which they should distinguish between. Ensure they realise that predators are consumers who hunt for, kill and eat other consumers. An animal is not a predator if it eats only plants! The prey is the animal which is eaten.
- Ask pupils to predict what would happen to the mouse population if there were no cats. All pupils would predict that it would go up. Conclude that predation is one factor which limits population size.

To make this clear, use a predator game as follows. Put some tables together to make a large surface. Position some prey (A5 sheets of paper) around the tables. Blindfold a child and give them one minute to find as many sheets as possible. They should only be allowed to tap fingers in different places on the table to find the prey. Ask other pupils what adaptations the person is using to find the prey. Vary the numbers of predator and prey to show the effect on the numbers of each surviving.

Assessing pupils' learning

- Lay out a circus of pictures of predators and prey. Pupils can identify which is which and could write down the adaptations each one demonstrates to their niche.
- Pupils should plot data showing the way in which predator and prey populations fluctuate (Figure 23.1).
- Less able pupils can simply explain why, when there are more predators, the prey population goes down (because more of them are eaten). More able pupils could explain the cyclic fluctuation in numbers (see below).

KEY STAGE 4 CONCEPTS

Predation

Recap the information from KS3. When defining predators and prey, you can re-introduce the words carnivore (eats only animals), omnivore (eats animals and plants) and herbivore (eats only plants).

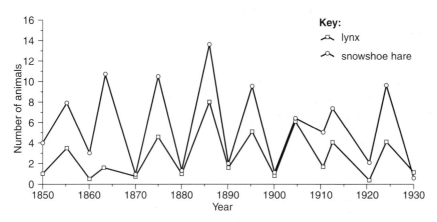

Figure 23.1 *The fluctuations in predator and prey populations. Data are taken from lynx and snowshoe hare populations in Canada*

There are a large range of adaptations shown by predators, e.g. teeth, webs, poisons, claws, group-hunting, camouflage. Either put out a circus of pictures of predators and prey, or ask pupils to research the adaptations of some named animals. Predator adaptations include good hearing, eyes facing forwards, fast running, large teeth, large claws, group hunting and poisons. The adaptations of prey may include good hearing, eyes positioned on the sides of the skull to give a greater field of view, fast running, camouflage and sharp horns for defence.

You can investigate the effect of camouflage on the likelihood of being predated. Make some round green and brown prey out of pastry (3 flour: 1 lard; 18 mm diameter and 1 mm depth) and food colouring. By pushing different coloured pastry together, also make some striped prey. Find a spot which you know is frequented by birds and ask pupils to distribute 10 of each kind. More able pupils could consider the importance of doing this randomly. For instance, if all the striped ones were placed under a bush, they may easily avoid being eaten. Return to count the number of prey of each type taken every 20 minutes. You should find that the brown and green prey will be predated more readily than the striped prey.

Reiterate from KS3 that predators can limit the size of the population of a prey species. In fact, predator and prey species' population size fluctuates cyclically. Extend the game from KS3 to help pupils understand such 'cyclic fluctuation'.

- Ask pupils what happens to the number of prey eaten when there are lots of predators. You should be able to demonstrate that if there are too many predators, they eat almost all the prey. It is then very difficult to find prey and predators will start to die off.

- Demonstrate that few prey are eaten when there are few predators. As a result, prey will survive to breed and their population size will increase. It is easier for predators to catch prey when the prey population size increases and the predators will again survive and reproduce.

With higher level pupils, you could talk through this process using a graph, with time on the X-axis and population size on the Y-axis (Figure 23.1). Use a different colour for predator and prey.

Assessing pupils' learning

- Ask pupils to identify adaptations of predators and prey.
- Pupils could design the perfect predator, or the perfect prey, taking features from a range of species and combining them into the same mythical species.
- Pupils could describe hunting or being hunted as if they were the predator or prey, highlighting adaptations of each.
- Pupils should plot data showing the way in which populations of predators and prey fluctuate cyclically. They should explain, in response to questions, why the predator and prey population fluctuate in this way.

chapter 24

Energy and nutrient transfer

BACKGROUND

Plants transform light energy into chemical energy and are the source of energy for all living things; plants are referred to as producers. Animals cannot use the Sun's energy; instead they gain a supply of chemical energy by eating plants or other animals; animals are consumers. The feeding, or trophic, relationships in a given habitat form food chains and food webs. A food chain describes one set of feeding relationships. For example, a lettuce (producer) may be eaten by a rabbit (primary consumer) which may be eaten by a fox (secondary consumer). A food web demonstrates how such food chains are interlinked, e.g. a snail may also eat the lettuce and a thrush may eat the snail. At each trophic level, some energy is used up before being transferred up to the next level. For this reason, the biomass that can be sustained at high trophic levels (e.g. tertiary consumer) is always smaller than that at low trophic levels (e.g. producer). We demonstrate this using pyramids of numbers and pyramids of biomass.

In contrast to energy, materials are recycled in food chains. Any waste or dead matter produced by a food chain is broken down by detritivores (worms and insects) and decomposers (fungi and bacteria). This material can then re-enter the chain by absorption into the producer through the roots (minerals and nitrogen) or the leaves (carbon dioxide). Recycling of carbon and nitrogen (the carbon cycle and the nitrogen cycle) are the examples of nutrient recycling included at KS3 and 4.

From KS2, pupils should know that food chains always begin with a green plant, although may not remember that green plants are first because only they can convert light energy to chemical energy. At KS3, introduce pupils to food webs, and pyramids of numbers and biomass. At KS4 double award, pupils must realise that food chains describe the flow of *energy* and materials through an ecosystem. They must be able to draw pyramids of numbers and biomass, and appreciate how decomposers are involved in recycling organic materials, including nitrogen and carbon.

KEY STAGE 3 CONCEPTS

Food chains and food webs

Ask pupils which is the only type of organism that can make its own food (which can convert light energy into chemical energy): plants. Most pupils should know that food is a store of energy (see Chapter 5). Because they produce a store of chemical energy from light energy, plants are called producers. To survive, animals also need (chemical) energy. They must consume energy from plants and other animals; they are therefore called consumers.

Many pupils arrive from KS2 concentrating on the fact that materials are transferred between organisms in a food chain. This is true, and you must reinforce it, but they must also realise that energy is transferred along a food chain. Many pupils fail to appreciate that without plants, no energy would enter the food chain and animals would not survive. Pupils sometimes believe that animals can use energy directly from the Sun, for example as heat. This is wrong: such energy does not help an animal move, excrete, reproduce etc. Its only effect may be to reduce the amount of chemical energy the animal expends keeping warm.

Ask pupils to build up a food chain on the board, and label the producers and consumers. Pupils find it easier to construct from the bottom upwards (i.e. beginning with plants and finishing with the top predator). Distinguish between primary consumers (animals which eat the producer), secondary consumers (animals which eat the primary consumer), and tertiary consumers (animals which eat the secondary consumer), and define the words herbivore (an animal which only eats plants), carnivore (an animal which only eats animals) and omnivore (an animal which eats both plants and animals).

Stress that the energy gained by carnivores does originate from a plant, because the animal which was eaten may have been a herbivore, or preyed upon a herbivore itself. Stress also that those animals high up in the food chain do not feed upon all other species in the food chain. They just feed upon that species immediately below them.

Many pupils draw the arrows on the food chain in the wrong direction. Arrows between species denote *energy* flow, and so should point *from* the producer *to* the top consumer. The arrows point into the mouth of the next consumer up the chain.

Science and Plants for Schools (SAPS: http://www-saps.plantsci.cam.ac.uk/) has a protocol for making an eco-column – essentially a stack of lemonade bottles containing a small-scale food chain which you can maintain in the laboratory.

Next, build up the idea of a food web. Ask pupils to go into the environment (perhaps the school field) and write down three food chains each. On the board, take one food chain from each pupil; the chains will interlink, building up a food web.

Pupils should appreciate the interdependence of populations within the food web. Ask them to predict what happens to the population of other species in a web, if the population size of one particular species were to increase. Ensure pupils predict appropriately for all food chains in the web, not just those in which that species is directly involved.

Assessing pupils' learning

- Within the context of named food chains, ask pupils to define key words such as producer, primary and secondary consumer, herbivore, carnivore and omnivore.
- Ask pupils to draw out food chains and webs based on information about feeding relationships.
- Watch a wildlife programme and ask pupils to describe a food web from the species featured. You could offer incentives for the largest food web described. You could do the same in class, but use books or CD-ROMs to gain the same information.

195

KEY STAGE 4 CONCEPTS

Food chains and food webs

Take your pupils outside and define the words community (a number of populations of different species) and environment (the place in which they live). The two together are called an ecosystem. Recap the work from KS3, particularly stressing that food chains represent the flow of energy. Pupils should know that the different feeding levels of the food chain (producer, primary consumer etc.) are *trophic* levels.

Assessing pupils' learning

Use similar activities to KS3.

KEY STAGE 3 CONCEPTS

Loss of energy from food chains

Ask pupils to think of ways in which energy is lost from a food chain. That is, why some energy is not passed up to a higher trophic level:

- Species at different feeding levels use up energy to survive (e.g. in movement and reproduction).
- Some energy contained within a species (e.g. in bones) is not available to species higher up the food chain (e.g. humans cannot digest bones and so cannot release energy from the chemicals stored within them).
- Some energy is wasted in excretion (e.g. urine, faeces, skin, hair).

> ### Assessing pupils' learning
>
> Ask pupils to add arrows to a food chain to denote the ways in which energy is lost.

KEY STAGE 4 CONCEPTS

Loss of energy from food chains

Remind pupils of how energy can be lost from a food chain. Ask them if more energy is lost from a food chain containing three members or from a food chain containing only two members. Most will answer that more energy is lost from a three-member food chain. Farmers can grow more food on land simply by harvesting the producer. If they harvest a primary consumer, it would already have lost some energy. If farmers harvested the producer, they would harvest that wasted energy while it was still contained within the crop.

Some of your pupils may be vegetarian for this reason: to use land to graze cattle rather than grow arable crops is sometimes seen as a waste of food production capability, preventing adequate food production for the world's population.

> ### Assessing pupils' learning
>
> Provide pupils with data about how much biomass is harvested by farmers from different levels of a food chain, and ask them to conclude why farming is more efficient when the producers are harvested.

Pyramids of numbers and biomass

Pyramids are used to show how energy is transferred between members of a food chain. There are two types of pyramids: numbers and biomass.

The length of the bars at each level of the pyramid of numbers should be directly proportional to the number of individuals of each species counted. Give pupils data for a particular food chain for which they must construct a pyramid of numbers. Ensure that the numbers of individuals at each trophic level decrease as you go up the food chain. This is because energy is lost at each trophic level; the amount of energy which can be transferred up to the next trophic level will always decrease and will sustain fewer individuals.

Pyramids of numbers do not always represent accurately what is happening in an ecosystem. Provide pupils with data for the following food chain and ask them to draw the pyramid of numbers:

Oak Tree → Caterpillar → Blackbird → Cat
(1)   (200)   (4)   (1)

The pyramid of numbers formed is inverted. If you ask pupils whether this is correct, most will answer no. However, to correct it, most pupils will suggest increasing the amount of oak trees to make a pyramid shape; a solution which suggests they do not understand the purpose of drawing pyramids (to represent actual energy flow happening in the ecosystem).

For less able pupils, make up two pyramids: (i) Use six books at the base, two on the next level up, and one on the top of the pyramid, and (ii) Use one book at the base, 40 sheets of A5 paper for the next level, and a pen for the final level. Both are upright pyramids, but if you draw a pyramid of numbers for each, the second one appears inverted. This is because individuals at each trophic level are not the same size.

You must remind pupils that the pyramid is there to demonstrate what happens to energy as it flows through a food chain. Remind them that energy enters the food chain in plants. Only some of this energy passes up to the next trophic level; hence, the next trophic level *must* in any food chain be smaller. If the pyramid is inverted, it does not represent energy flow properly.

Ask more able pupils how to correct this. Some pupils will suggest counting the leaves on the tree, and re-drawing the pyramid of numbers. More able pupils will realise that the pyramid is inverted because the tree and the caterpillars have such different masses. Construct a pyramid of biomass, where the area of the bars represents the total biomass present at each trophic level. Pyramids of biomass always yield an upright pyramid.

- Pupils should identify ways in which energy is lost in named food chains.
- Pupils should practise constructing pyramids of number and biomass.
- More able pupils could explain what is wrong with using a pyramid of numbers to estimate energy flow in a food chain.

KEY STAGE 4 CONCEPTS

Pyramids of numbers and biomass

Remind pupils that pyramids of numbers represent energy flow and that energy is lost from food chains at each trophic level for the reasons outlined at KS3. Give pupils practice drawing pyramids. Present them with an inverted pyramid of numbers again and ask them to explain the problem. You may need to help less able pupils to realise the problem.

At this level, pupils should realise that there must be more *energy* at a lower trophic level to feed the next trophic level. Explain this in two ways:

- Tell less able pupils that each unit of biomass can only survive if it is provided with a proportionate amount of energy. Therefore, biomass is correlated with energy content.
- More able pupils will realise that living organisms are constructed from atoms linked by bonds. Although not strictly correct, biologists often say that energy is stored in those bonds. More able pupils will realise that the more biomass, the more bonds, and therefore the more energy.

Whichever explanation you use, by estimating energy using mass instead of numbers, we gain a better estimate of the energy at each trophic level. Because we are using pyramids to represent energy flow, this is the best pyramid to use. Stress that biomass should be dry mass (i.e. with all the water removed). This is because living things cannot access the energy stored in the bonds within water.

Assessing pupils' learning

- Provide practice questions about pyramids of numbers and biomass.
- Ask more able pupils to explain why biomass is a better approximation of energy content at each trophic level than numbers.
- More able pupils may also be able to explain why we measure the dry mass.

198

Recycling materials from a food chain

Begin by asking why humans recycle: why don't we simply pile up all our tin cans? The answer is that we would run out of raw materials to make new tin cans. Ask pupils what happens to the waste and dead matter which originates from food chains. Some will say it disappears, implying they do not understand that matter is conserved. Most should predict that it rots or decays. They may not appreciate that when it rots, it is returned to the soil and broken down into its constituent chemicals (i.e. to the abiotic environment); it can then re-enter the food chain.

In general, soft body parts will rot most quickly and hard body parts will rot most slowly.

- Compile a list of household waste, including food, paper, plastics and metals. Alternatively, bring in the contents of your dustbin. Pupils should decide which types of waste will decompose quickly, slowly, or not at all. They should realise that only biological material will decompose, and the softest material will decompose most quickly.
- *Science and Plants for Schools* (SAPS: http://www-saps.plantsci.cam.ac.uk/) has a method to make a compost column out of old lemonade bottles. This allows less able pupils to investigate which materials will or will not decompose.
- You can find the mass of rotting organic matter in the soil by burning. Dry a soil sample in an oven (50°C) for three days to remove the water. Find the mass of the soil sample, heat the sample with intense heat for five minutes, and find the mass again. The loss in mass represents the amount of organic matter in the sample.

Safety Advice: Ensure pupils wash their hands thoroughly after each practical. Only handle the dustbin contents yourself and wear gloves.

At this point, introduce the idea that rotting is caused by organisms. These include detritivores (e.g. worms eat detritus (waste material) and break it down) and decomposers (e.g. fungi secrete enzymes onto waste material to digest food which is then absorbed). The effect of this mechanical and chemical breakdown makes soft matter become mushy and rotten. More able pupils may realise that it is more difficult for enzymes to penetrate harder organic material and so those decompose more slowly. Because enzymes are adapted to break down organic matter, they have no effect on man-made household waste.

Pupils must appreciate the conditions required for decay to be successful. Ask pupils what is required for bacteria and fungi to survive.

- Like any living thing, oxygen is needed for respiration.
- Decay will move faster if there are more decomposers. Bacteria will reproduce more quickly if it is warm.

- Like every organism, decomposers need water, both for normal life processes, to secrete solutions of digestive enzymes and to absorb the products of digestion.

Ask pupils to investigate the effect of temperature and/or moisture on the rate of decay of peas in a film case using a fridge, freezer, room temperature and an incubator (Figure 24.1). Keep the lid on the cases or the water will evaporate. Do not let pupils put the peas under the cotton wool: they will not have access to oxygen.

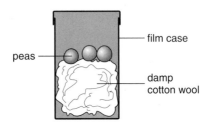

Figure 24.1 *Apparatus to investigate the factors which affect the decay of food. Vary the temperature at which you keep the film cases, or the moisture content of the cotton wool*

S Safety Advice:

- Advise caution when opening the lids of film cases containing decayed peas. Spores from the decomposers could be inhaled or ingested.
- Do not incubate peas above 30°C. This should avoid culturing pathogenic bacteria which would normally thrive at human body temperature (37°C).

Assessing pupils' learning

- Ask pupils to make a list of all the sources of waste material from a food chain. Some examples include dead animals, dead plants, urine, faeces, hair, leaves, fur and skin.
- Pupils should explain why soft organic materials will decompose more readily than hard organic materials. They should be aware, and more able pupils should explain, why inorganic materials will not decompose.
- Provide pupils with some foods preserved in different ways (e.g. ultra heat-treated milk, refrigerated fresh milk, dried coffee, pickled beetroot, jam, salted ham). Ask them to describe which condition required for rotting was unavailable. Remember, the ethanoic acid in pickling vinegar, the high sugar concentration in jam and the salt in salted ham will actually kill the decomposers. More able pupils may realise that the sugar and salt

will cause water to pass out of decomposers by osmosis: they will then collapse and die.
- The film case experiment should be written up. Pupils should explain why decomposers need moisture, warmth and oxygen to decay food effectively.

The carbon cycle

Remind pupils that matter is recycled, not destroyed, between the biotic and abiotic components of the environment. The carbon cycle itself is difficult to make interesting. You can find a diagram in any textbook. It helps if pupils realise why carbon is so important. Carbon is a major structural component of our bodies. Proteins, carbohydrates and fats are all made with carbon, and all living organisms are made of proteins, fats and carbohydrates. If you did not recycle carbon, there would be lots of carcasses of carbon, and new living things would not receive a supply of carbon from which their new bodies could be constructed.

To introduce the carbon cycle, you can do two things. First, you could try and build up the cycle on the board from your pupils' ideas. Ensure you have *your* cycle written out in front of you; it is very easy to draw a cycle which ends up being a mass of criss-crossing arrows, and appears very confusing. A safer bet may be to draw out the cycle on an overhead transparency and reveal it bit-by-bit as your pupils think of ideas. More able pupils could conduct their own research project on the carbon cycle.

Assessing pupils' learning

- Higher level pupils could draw out their own cycles, perhaps with the aid of a textbook. Less able pupils could complete a cloze-style exercise with the same aim. Ask all pupils to include upon their cycle the location of decay, and the organisms involved in that decay.
- Less able pupils could produce a poster or a 'mobile' of the carbon cycle.
- Ask pupils to use their cycles to describe the flow of carbon between particular points of the cycle. Pupils could imagine themselves to be carbon atoms, describing their journey around the carbon cycle.
- Ask pupils to write a poem, outlining the main stages of the cycle.

The nitrogen cycle

In many syllabuses, only higher level pupils must be familiar with the nitrogen cycle. You could introduce it in a similar way to the carbon cycle. A diagram will be in any textbook. Again, begin by asking why organisms need nitrogen: it is used to make protein. Proteins make enzymes and are used for growth and repair.

Tell pupils that, unlike carbon dioxide, plants have no method of getting nitrogen into the food chain themselves, even though nitrogen makes up 79% of the air. Because of this, they must take in nitrates. These are produced in three ways:

- by nitrogen-fixing bacteria in the soil and in nodules (lumps) on the roots of leguminous plants (e.g. peas, beans and clover).
- by nitrifying bacteria which break down dead and waste organic matter in the soil into nitrates.
- by lightning, whose high energy causes nitrogen in the air to react with oxygen and form nitrates.

Nitrates can also be added to the soil in the form of fertilisers.

You can demonstrate the presence of nitrogen in soil using nitrate and ammonium test strips. These are sold at most large pet shops for testing aquarium water. Simply shake up a soil sample with distilled water and test the water with the test strip.

S **Safety Advice:** Be aware that there is a risk of tetanus infection from taking soil samples. Cover open cuts and wash hands thoroughly afterwards.

Having summarised this, explain how nitrates are lost from the soil:

- by denitrifying bacteria converting nitrate back to nitrogen gas.
- by plants using nitrates to make proteins which are then eaten by animals.
- by leaching of excess nitrates into rivers.

Assessing pupils' learning

- Pupils could draw out their own cycles, or complete a cloze-style exercise.
- Ask pupils questions on the flow of nitrogen around parts of the cycle. They should be able to explain, for instance, how nitrogen is obtained by animals, or how nitrogen is lost from the soil. Again, they could do so by imagining themselves to be nitrogen atoms, describing their journey around the cycle.
- Less able pupils could produce a poster or 'mobile' of the nitrogen cycle.
- The most able pupils could research parts of the nitrogen cycle for homework and write in detail about the chemical processes happening at each stage.

chapter 25

Human impact on the environment

BACKGROUND

The human population is growing exponentially. Such population growth has caused an increased consumption of natural resources and an increased production of waste and pollution. Both of these factors can have detrimental environmental effects.

Because the human population size is growing, farming has had to become more efficient. Intensive farming methods are designed to maximise production at whichever level of the food chain a crop is harvested. For example, farmers may add nitrogen fertilisers to help plant growth, but these can pollute rivers and lakes causing eutrophication. Pesticides, including herbicides (weed-killers), insecticides (insect-killers) and fungicides (fungus-killers), are also added to crops. Using herbicides prevents weeds competing with crop plants for sunlight, minerals and water. Insecticides and fungicides prevent energy being lost from the crop plant to insects and fungi, respectively. There are specific environmental problems caused by each of these chemicals. They can all accumulate in food chains, poisoning consumers at high trophic levels.

Pupils from KS2 will not have considered man's effect on the environment. At KS3, pupils should understand that toxic pesticides will accumulate in food chains, poisoning those species at high trophic levels. At KS4 double and single award, pupils must appreciate the reasons for intensive farming methods, and appreciate their environmental effects. The effect of an increasing human population on natural resource consumption and pollution should also be understood. Research suggests that environmental education in schools can feed back to parents and positively influence their behaviour, for instance in recycling. Bear this in mind and reinforce to pupils that each of them has a responsibility to their environment.

KEY STAGE 4 CONCEPTS

The increasing human population

Begin by providing students with data for human population size over the last 2000 years. When they plot the data against time, they will see that human population size is increasing exponentially. In the last 200 years, population size has more than doubled. This is a greater increase than that which occurred between the years 0 and 1800.

> **Assessing pupils' learning**
> - Ask pupils to think of reasons for this sudden increase in population size: better health care, technology and food production.
> - Ask pupils to think about the consequences of a suddenly larger population: they could focus on the use of finite resources (e.g. minerals, natural resources and fossil fuels) and the production of harmful waste (household waste, carbon dioxide, nitrogen oxide, sulphur dioxide and sewage).
> - Use questions to check pupils understand that the increasing gradient of the graph means population size is increasing more and more quickly.
> - Ask pupils to produce a display showing the reasons for and consequences of this increase.

KEY STAGE 3 CONCEPTS

Increased food production

At KS3, pupils do not need to know in detail why farmers use artificial chemicals to aid food production. However, they must appreciate how such chemicals can poison animals high up the food chain. For more able pupils, you could use a historical example of DDT to explain this (this pesticide is the most usual example used by biology teachers and its story is included in most textbooks). For less able pupils, use the following example and role play:

e.g.

$$\text{Cabbage} \rightarrow \text{Butterfly} \rightarrow \text{Sparrow} \rightarrow \text{Hawk}$$

Label some pupils as butterflies, less as sparrows and still less pupils as hawks. Give each butterfly one ball (to represent the toxin). Allow predation to occur (perhaps through a game of tag on the school field) and eventually all the balls will be held by the hawk. Hence the toxin will have accumulated in the hawk.

The key biological points to stress include:

- The toxins (e.g. DDT) normally used to kill pests (e.g. aphids) may be absorbed by other species (e.g. butterflies) without killing them.
- Each such butterfly would be eaten by a sparrow who would ingest many doses of toxin by eating lots of butterflies. The sparrows cannot excrete the toxin or break it down; it accumulates in the fat reserves in their bodies.
- When the hawk eats many such sparrows, the toxin becomes concentrated so much in its body that it has catastrophic effects on the hawk population.

Assessing pupils' learning

- Give pupils some numerical questions about how toxins accumulate in food chains.
- Ask pupils to produce a flow chart, showing how toxins accumulate in food chains.

KEY STAGE 4 CONCEPTS

Increased food production

Begin by recapping ways in which energy is lost from food chains (see Chapter 24). Farmers have a variety of methods to minimise the energy lost from a food chain, and therefore to maximise the amount of crop which can be harvested. A variety of these and their effects on the environment are outlined below.

Battery farming
Battery farming is practised to minimise energy loss from the food chain. Ask pupils to consider how chickens save energy by living in cramped conditions with food supplied, and a constant temperature maintained in the sheds. Possible ideas include: less energy wasted in movement, less energy wasted finding food, less energy wasted keeping warm, no predators. All of these allow animals to invest more energy into producing eggs. More able students could consider the ethical implications of battery farming. The RSPCA produces leaflets and booklets about animal welfare which may aid discussion.

Fertilisers
Fertilisers contain extra minerals which are needed by the plant (see Chapter 15). Unfortunately, excess fertilisers can be washed out of the soil by rain water into lakes and rivers (this process is called leaching). Fertilisers in rivers can cause eutrophication. Pupils find this process difficult to understand and remember.

The details of the process are outlined below; try to explain it in the form of a flow chart which you build up with a series of questions. Less able students are unlikely to require the detail of this process.

- Ask pupils whether fertilisers in rivers will help growth of water plants, as they help growth of land plants. Most pupils will answer that they will. To confirm this, you could repeat the experiments from Chapter 15 about the effects of fertiliser on duckweed.
- In response to fertiliser, algae reproduce in large numbers, and form a thick layer on top of the water which blocks light getting through to plants living on the river bed.
- If these plants receive no light, they will die because they cannot photosynthesise. Bacteria then decompose the dead plants (and some of the algae which also die, having run out of nutrients from the fertiliser). Because these bacteria respire aerobically (see Chapter 5), they use up all the oxygen in the water. This lack of oxygen causes fish and other animals also to die.

Pesticides

Ask pupils to consider how the use of pesticides and insecticides can minimise energy loss from a food chain. They could research the negative effects of fungicides, herbicides and insecticides on their environment.

> **Assessing pupils' learning**
> - Small groups could research one of the above topics and give a presentation to the rest of the class.
> - Provide pupils with a map of a farm and data about how much fertiliser, insecticide and herbicide has been added to each field. Give them a list of written outcomes (e.g. eutrophication occurs, an endangered species of insect becomes extinct etc.). Ask them to match these outcomes to specific fields, or parts of a river.

The effects of increasing human population

Sewage

Research has shown that pupils think that anything natural or biodegradable is not pollution, and does not have harmful effects. That is wrong: the production of more sewage by an increasing human population leads to higher levels of nitrates put onto land and into rivers. The efflux of nitrogen-rich compounds from sewage works into rivers could cause eutrophication, as outlined above. More able students could consider alternative disposal methods. Disposal at sea is about to be banned, and disposal as manure on farms is unpopular because of a perceived risk to human health posed by contaminated crops.

Fossil fuels

Burning of fossil fuels, in the form of coal, natural gas, or oil products, releases smoke particles, sulphur dioxide and carbon dioxide. Demonstrate this by heating coal (Figure 25.1). Replace the indicator with limewater to show the presence of carbon dioxide.

Safety Advice: Be careful of 'suck-back' when you finish heating.

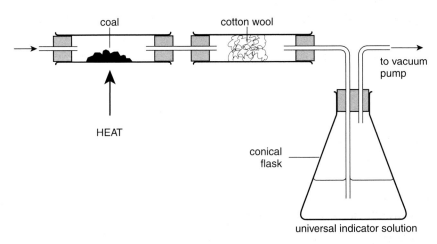

Figure 25.1 *Apparatus to demonstrate pollution caused by burning of coal. The cotton wool will discolour and the universal indicator turn red, indicating the production of acidic gases*

Smoke particles can settle out of the atmosphere as dust, causing respiratory diseases. Dust can also block the stomata of plants, and covers their surfaces, inhibiting photosynthesis (see Chapter 15). Pupils can observe such pollution by (i) washing leaves in water and filtering the washings, (ii) sticking Sellotape to a leaf, peeling it off and sticking it to white paper, or (iii) pressing slightly damp tissue against the surfaces of a leaf for several hours, perhaps under a pile of books, and then examining the tissue. Choosing leaves from a roadside verge will give you the best results, but be aware of road safety.

Sulphur dioxide produced from the burning of fossil fuels dissolves in water vapour in the air, producing sulphuric acid ('acid rain'). When it rains, this can damage plants. There are four practicals which demonstrate the effect of acid rain.

1 Set up the apparatus shown in Figure 25.2. The sodium metabisulphite gives off sulphur dioxide. If different groups of pupils set up one concentration of sodium metabisulphite each, you should demonstrate that acid rain inhibits germination of wheat seeds.

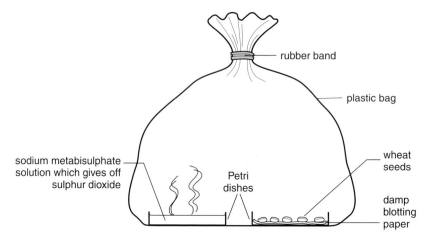

Figure 25.2 *Investigating the effect of acid rain on germination of wheat seeds*

2 Cress seeds 'watered' on a Petri dish with solutions of increasingly strong sulphuric acid (0.001 M, 0.01 M, 0.1 M, and 1 M) should germinate less and less quickly.

3 Place one leaf into a container of water and one into a container of vinegar. Put a lid on the containers and leave overnight. After 24 hours, the leaf will have developed black spots, or will have turned brown, depending on the species.

4 Ask pupils to investigate the effect of acid rain on limestone buildings and chalk cliffs. Give them marble chips and 1 M sulphuric acid. Pupils should have met reactions of acid and metal carbonates in KS3 chemistry.

S **Safety Advice:** Ensure safety glasses are used for all four practicals.

Production of carbon dioxide by burning also unbalances the carbon cycle, leading, as we produce more and more carbon dioxide gas, to global warming. Lenton & Lenton (1998) describe a short practical to quantify the greenhouse gases produced in exhaust fumes. You should explain the basic principle of how the greenhouse effect occurs; long wave radiation incident on the Earth passes unhindered through the atmosphere, but some of the short wave radiation re-radiated from the Earth is trapped by 'so-called' greenhouse gases.

Be aware that many pupils confuse global warming and ozone hole depletion (Potts *et al.* 1996). Research demonstrated that 21% of pupils at KS3 think carbon dioxide can damage the ozone layer. Remember, pollutants such as chlorofluorocarbons (CFCs) used in aerosols and refrigerators destroy ozone which normally absorbs harmful ultra-violet rays from the Sun. Asking pupils to explicitly compare the two processes is useful for destroying such misconceptions.

Minerals

Supplies of minerals are limited. If we continue to extract them from the ground, we will eventually run out. We are also left with metal and plastic products which are not biodegradable and are placed into land-fill sites. Production of such sites disrupts natural habitats, and may put toxins into the environment. Because of this, we are encouraged to recycle valuable minerals.

Pupils could conduct a rubbish survey at home, producing a report which outlines how much rubbish of different types is produced in their house on a typical day. They could also make recommendations about what should be sent for recycling rather than being consigned to the rubbish bin.

Safety Advice: Warn pupils about the health risks of going through the rubbish bin. They should ask their family to keep a record every time something is thrown away. If they must go through the bin, they should ask parental permission and wash their hands thoroughly afterwards.

Assessing pupils' learning

- Pupils should show written understanding of each type of pollution, either through notes or a series of structured questions.
- They could consider the financial and political implications for minimising such pollution. SATIS produce a variety of resources for classroom debates which stress both the scientific, political and financial implications of pollution and recycling.
- Pupils could produce leaflets or posters outlining the major causes of pollution. More able pupils could investigate the chemical reactions involved in the formation of acid rain.
- Pupils should write letters to the local council, explaining where they think there is a serious pollution problem in the neighbourhood, and outline its potential effects on the environment.

References

Horsley, A. (1990) Acid rain in the lab: an investigation into the effects of sulphur dioxide gas on wheat seed germination. *Journal of Biological Education* 24:71

Perkins, P. (1993) The effect of sulphur dioxide on the germination and growth of cress seeds. *School Science Review* 72:90–92

Potts, A., Stanisstreet, M. & Boyes, E. (1996) Childrens' ideas about the ozone layer and opportunities for physics teaching. *School Science Review* 78:57–62

PART ONE

KS3

1 The drawing shows part of an organ system.

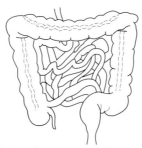

a) Which organ system is the diagram part of?

circulatory system	☐	reproductive system	☐
digestive system	☑	respiratory system	☐

1 mark

b) Which line indicates most accurately where this part of the organ system is found in the body?

1 mark

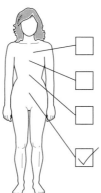

2 Below is a drawing of a leaf cell from a plant.

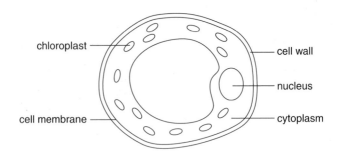

a) i) Which part of this cell is *not* present in a root cell?

chloroplast

1 mark

ii) Which two parts of this cell are *not* present in an animal.

1 *chloroplast*　　　　　　2 *cell wall*

2 marks

b) In the table below, write the correct part of the cell next to its function. The first one has been done for you.

Function	Part of the cell
A place where many chemical reactions take place	Cytoplasm
Photosynthesis takes place here	*chloroplast*
It controls the cell's activities	*nucleus*
It helps to keep the shape of the cell	*cell wall*
It controls substances entering and leaving the cell	*cell membrane*

4 marks

KS4

1 When Joy took her driving test she was asked to do an emergency stop. The test examiner slapped his hand on the front of the car. This was a signal for Joy to brake hard. Her reaction time was 0.5 seconds.

a) Which of Joy's senses would pick up the examiner's signal?

hearing

1 mark

b) Which of the following types of cell, **A, B, C, D** or **E**, would carry the message from the sense organ to the brain?

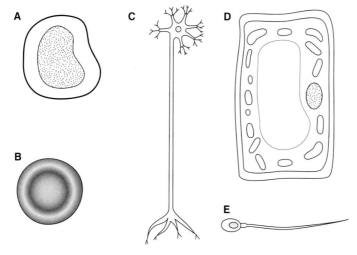

Answer = *C*

....

1 mark

c) Choose words from this list to answer the questions.

digestive   excretory   hormonal   nervous   respiratory

i)   Name the system in the human body in which cells carrying messages
are found.

Answer = *nervous*

....

1 mark

ii)   Name the *other* system in the human body which allows us to respond
to changes.

Answer = *hormonal*

....

1 mark

d) The test examiner decided to find out about reaction times. He measured
and recorded the reaction times of other drivers. The results are shown in the
table and some have been shown on the bar chart. Use the results to
complete the bar chart.

Table of reaction times

Reaction time in seconds	Number of drivers
0.4	2
0.5	6
0.6	9
0.7	6
0.8	3

Bar chart of reaction times

Bar chart completed: 0.5 s = 6 drivers, 0.8 s = 3 drivers   *2 marks*

e) Use the results to help you to answer these questions:

i) How many drivers had a reaction time of 0.5 seconds?

6 .... *1 mark*

ii) Which reaction time was the most common?

0.6 seconds .... *1 mark*

iii) Describe the pattern shown by the results.

Few people have very slow or very quick reaction times. Most people

have a reaction time of 0.6 s. .... *2 marks*

f) Drivers are told not to drink alcohol and drive.

i) If drivers do drink alcohol how will their reaction times change?

increase .... *1 mark*

ii) Name *one* other thing which can change a driver's reaction time.

tiredness/other drugs .... *1 mark*

OCR

2 The diagram below shows a cell containing the diploid number of chromosomes.

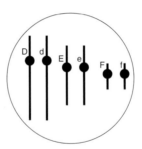

One pair of alleles on each pair of chromosomes has been shown.

a) How many genetically *different* types of gamete can be produced following meiosis of this cell? Tick the correct box below to show your answer.

Number of different types of gamete	Tick
3	
4	
6	
8	✓

1 mark

b) *Three* of the following diagrams (**A**, **B**, **C**, **D**, **E** and **F**) show stages in meiosis in this cell.

Put a tick (✓) in the box beside each diagram which shows a stage in meiosis in this cell.

3 marks

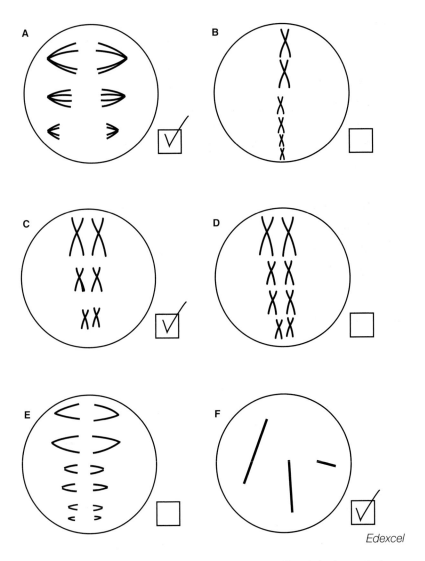

Edexcel

3 This question is about support in plant cells. The plant cells are in the root of a carrot.

The diagram shows three carrot root cells placed in a concentrated sugar solution.

a) i) Name the parts of the cell labelled **A**, **B** and **C**. Write the names in the boxes.

Part of cell	Name
A	cell wall
B	cell membrane
C	cytoplasm

3 marks

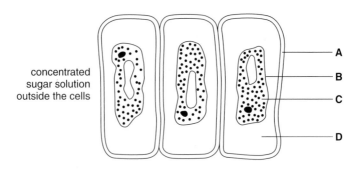

concentrated
sugar solution
outside the cells

A

B

C

D

ii) What is filling the space labelled **D** on the diagram?

sugar solution from outside the cell

1 mark

b) The boxes have descriptions of plant cells which have turgor pressure and plant cells which do not. Tick the *three* boxes that contain descriptions of cells with turgor pressure.

the cell membrane is pressed against the cell wall	☑
the cell membrane is not in contact with the cell wall	☐
the cell wall is stretched	☑
the cell wall is not stretched	☐
the vacuole is pushing against the cytoplasm	☑
the vacuole is not pushing against the cytoplasm	☐

3 marks

c) People do not like buying carrots which are split. Scientists can use a wedge test to find out how easily a carrot splits. The diagram shows a wedge test.

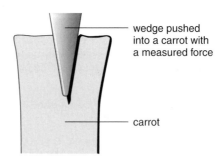

wedge pushed
into a carrot with
a measured force

carrot

The graph shows the results of many wedge tests.

i) What does the graph tell you about the relationship between turgor pressure and the amount of force needed to split carrot roots?

as turgor pressure increases, the amount of force required decreases

1 mark

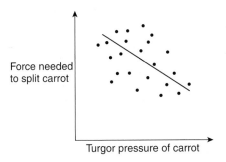

Force needed to split carrot

Turgor pressure of carrot

ii) Suggest why carrots grown in low-rainfall areas are less likely to split.

because they receive less water, their cells will have less turgor pressure

1 mark

OCR

PART TWO

KS3

1 The table shows a copy of the nutritional information from the label of a can of baked beans.

Nutritional information	List of ingredients	
100 g provides:		
energy	406 kJ	
protein	5.4 g	haricot beans, tomato purée,
total carbohydrate	17.6 g	water, sugar, modified starch,
sugar	6.0 g	salt, paprika, onion powder,
fat	0.4 g	herb extracts, spices
fibre	3.7 g	

a) A balanced diet contains several different groups of substances. Give *one* substance required for a balanced diet, which is missing from the above information.

vitamins or minerals

1 mark

b) i) Choose the food, from the list of ingredients, which contains the most protein.

haricot beans

1 mark

ii) Why do we need protein in our diet? Give one reason.

growth/repair/making proteins/replacement of cells

1 mark

c) i) Choose the food, from the list of ingredients, which provides fibre.

haricot beans/tomato purée

1 mark

ii) Fibre is needed for a balanced diet. Why?

helps the passage of food through the gut/prevents constipation

1 mark

2 a) Below is a diagram showing the flow of blood to and from a muscle.
i) Name the type of blood vessel labelled B.

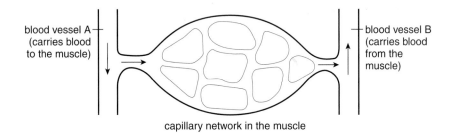

blood vessel A
(carries blood
to the muscle)

blood vessel B
(carries blood
from the
muscle)

capillary network in the muscle

vein

1 mark

ii) Respiration occurs in muscle cells. Use the table to compare oxygen, glucose and carbon dioxide concentrations in blood vessels A and B. Place a tick in one box in each row of the table.

Substance	A higher concentration in A than in B	The same concentration in A and B	A lower concentration in A than in B
Oxygen	✓		
Glucose	✓		
Carbon dioxide			✓

3 marks

b) Blood flows to and from each organ in the body.
i) Which organ is represented by the diagram below?

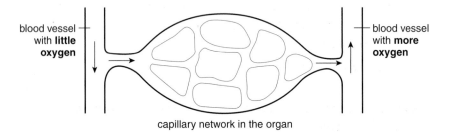

blood vessel
with **little**
oxygen

blood vessel
with **more**
oxygen

capillary network in the organ

lungs ....

1 mark

ii) The diagram below is of a different organ. Which organ is represented?

small intestine ....

1 mark

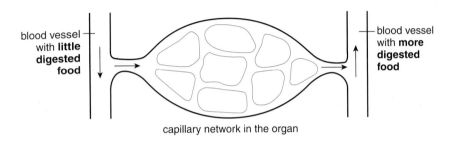

blood vessel with **little** digested food

blood vessel with **more** digested food

capillary network in the organ

c) If someone has an alcoholic drink, alcohol will pass into the blood and be circulated around the body.

i) Blood vessels in the skin become wider when alcohol is in the blood. What is the effect of this on the body?

cools the body/skin feels warm/red skin ....

1 mark

ii) Tick the correct box to show which of the following is a long-term effect of alcohol abuse.

Tick the correct box.

shorter reaction time	☐	short sightedness	☐
damage to liver	☑	damage to bones	☐

1 mark

3 When writing to a new pen-friend, Robert described himself as follows:

I am male.	**My weight is 600 N.**
I am 17 years old.	**I can speak French.**
My eyes are brown.	**I have a scar on my chin.**
My height is 1.8 m.	

a) Choose *two* characteristics which must have been inherited and have therefore not been affected by his environment.

1 _I am male_ ....

2 _My eyes are brown_ ....

2 marks

b) Choose *two* chracteristics which may be affected by both his environment and by inheritance.

1   *My height is 1.8 m* ....

2   *My weight is 600 N* ....

2 marks

c) Robert measured the height of 15-year-old children in his scout group. He used a bar chart to record his results.

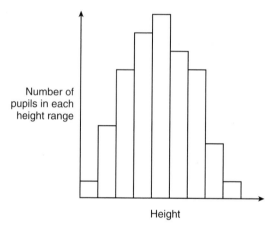

Number of pupils in each height range

Height

He made a list of features about which to collect data on another occasion. Which two of those features in the list would have a distribution similar to Robert's bar chart?

ability to roll the tongue   ☐   circumference of the head   ☑

presence of ear lobes   ☐   sex of the pupil   ☐

mass of the pupil   ☑

2 marks

4 a) Diagram 1 shows the muscles and bones in a leg.

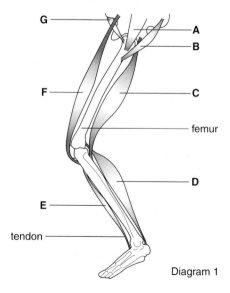

G

A

B

F

C

femur

D

E

tendon

Diagram 1

i) If muscle **A** contracts. What effect does it have on the leg?

pulls the leg to the side

1 mark

ii) Write down the letters of the antagonistic pair of muscles which control the bending and extending of the leg at the knee.

C and F

1 mark

iii) In an antagonistic pair, when one muscle contracts, the other does not completely relax. What is the advantage ot the other muscle maintaining some tension?

it allows controlled or slow movements

it allows small movements

1 mark

b) Diagram 2 below shows the elbow joint. The ends of the bones at the joint are covered by tissue **X**

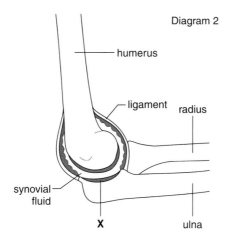

Diagram 2

humerus

ligament

radius

synovial fluid

X

ulna

221

i) What is the name of tissue **X**?

Cartilage

1 mark

ii) People with osteoarthritis have joints in which small pieces of tissue **X** break off. What two effects may this have on the joint?

Ends of bones will rub together

Less smooth movement/stiffness inflammation or swelling

2 marks

c) Diagram 1 shows tendons. Diagram 2 shows ligaments. Tendons stretch less than ligaments.

i) Why do the ligaments at the elbow need to stretch?

to allow the arm to bend

1 mark

ii) Why do tendons in joints hardly stretch at all?

if they stretched, they would not pull as much on the bones

1 mark

5 The diagram below shows the outside of a red blood cell.

a) Three characteristics of red blood cells are given below. Explain how each characteristic helps the cell to transport or exchange oxygen.

i) Haemoglobin is contained within red blood cells.

haemoglobin binds with oxygen

1 mark

ii) Red blood cells have a biconcave shape.

Large surface area for exchange of oxygen

More haemoglobin near the surface

1 mark

iii) Red blood cells have flexible membranes, enabling them to change shape.

allows red blood cells to squeeze through narrow capillaries

1 mark

b) Which metal is contained within the haemoglobin molecule?

iron

1 mark

c) Marathon runners often train at high altitude before a race. This increases the number of red blood cells in their blood. Why would a larger number of red blood cells help an athlete during a race?

More oxygen carried to the muscles

Energy released more quickly by respiration

Carbon dioxide removed from muscles faster

Cuts down build up of lactic acid

2 marks

d) The circulatory system carries oxygen from the lungs to the muscles. Which of the following flow charts shows the correct route taken by the oxygenated blood between leaving the lungs and arriving at the muscles?

pulmonary artery → left atrium → left ventricle → aorta ☑

pulmonary artery → right atrium → right ventricle → pulmonary vein ☐

pulmonary vein → left atrium → left ventricle → aorta ☐

pulmonary vein → right atrium → right ventricle → pulmonary artery ☐

6 The condition called emphysema damages air sacs in the lungs. The diagram below shows a damaged air sac and a normal air sac

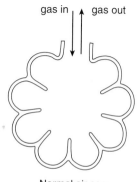

Normal air sac

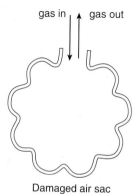

Damaged air sac

a) When a person breathes, exchange of gases occurs at the inside surface of the air sac.

i) Which two gases are exchanged in the air sac?

oxygen and carbon dioxide

1 mark

ii) Why is the amount of gas exchanged in a damaged air sac smaller than in a normal air sac?

Smaller surface area available

1 mark

b) The four substances below are present in cigarette smoke.

carbon particles   carbon monoxide   nicotine   tar

From the list choose the substance which:

i) causes addiction to cigarette smoking;

nicotine

1 mark

ii) can cause lung cancer;

tar ....

1 mark

iii) is carried instead of oxygen in the red blood cells.

carbon monoxide ....

1 mark

7 The diagram below represents a baby developing inside its mother.

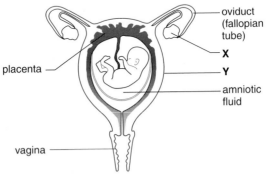

a) Organ **X** produces eggs. Name organ **X**.

ovary ....

1 mark

b) The baby develops in a bag of amniotic fluid. The bag of fluid sits in organ
Y. What is organ **Y**?

uterus or womb ....

1 mark

c) i) Which labelled organ in the diagram passes food from the mother to the
baby?

placenta ....

1 mark

ii) Which useful substance, other than food, can pass from the mother to
the baby?

oxygen ....

1 mark

d) What is the name of the organ system depicted in this diagram?

reproductive system ....

1 mark

1 The diagram shows part of the digestive system.

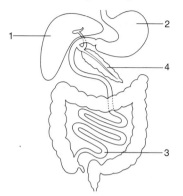

gastric juice   intestinal juice   pancreatic juice   bile

a) Match the chemical to the organ which secretes it.

Organ	Chemical
1	*bile*
2	*gastric juice*
3	*intestinal juice*
4	*pancreatic juice*

3 marks

b) The following nutritional information is found on a bag of crisps.

Food content	g in 100 g
Protein	5
Carbohydrate	60
Fat	35

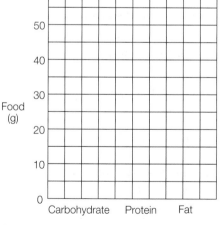

i) Draw a *bar chart* to show this information.

two bars to correct height = 1 mark
....

all bars to correct height = 2 marks
....

2 marks

ii) Name *one* other substance needed in a balanced diet.

vitamins/minerals/fibre

1 mark

c) The diagram shows part of the small intestine.

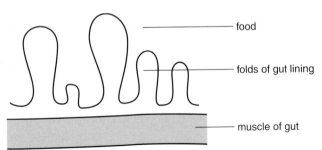

— food

— folds of gut lining

— muscle of gut

The folds increase the surface area touching the digested food. These folds have finger-like projections on them.

i) What are these projections called?

villi

1 mark

ii) What is the function of the small intestine?

digest and absorb food molecules

1 mark

d) A student took some vegetable oil, placed it on a microscope slide and observed diagram **A**.

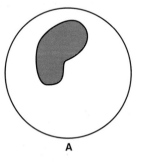
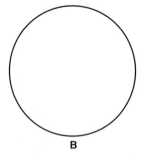

A

B

He then added some bile which breaks up the vegetable oil into tiny droplets.

i) In the circle, **B**, draw what the student saw after he added the bile.

oil spread over the surface of the slide in small droplets

1 mark

ii) What happened to the surface area of the vegetable oil when the student added the bile?

surface area increased

1 mark

e) Glucose molecules are so small that when you suck a sweet they pass from your mouth, through the lining of your cheek into the bloodstream.

i) In what part of the blood is the glucose carried?

plasma

1 mark

The glucose may be used to move the muscle of your arm in a process called respiration.

ii) Complete the equation to show how this process takes place.

Glucose + *Oxygen*

⟶ water + *carbon dioxide*

➤ Movement

2 marks

2 a) The sentences are about breathing. Choose words from the list to complete the sentences that follow.

Each word may be used once or not at all.

decreases diaphragm in increases intercostal lungs out ribs ventilation

Air is drawn into the *lungs* as the thorax *increases* in volume.
This change in volume is caused by the *intercostal* muscles contracting and moving the rib cage *out* at the same time as the *diaphragm* is moved down by its muscles.

5 marks

b) The diagram shows a part of a lung that is involved in gaseous exchange in a human.

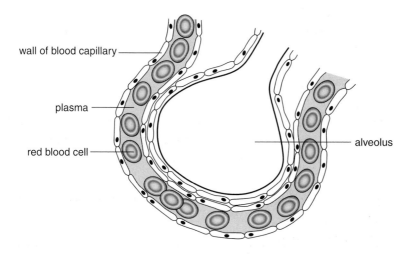

wall of blood capillary

plasma

red blood cell

alveolus

i) Draw and label, on the diagram, *one* arrow to show the direction of movement of oxygen between the alveolus and capillary.

arrow from air space in alveolus to red blood cell in capillary

1 mark

ii) Draw and label, on the diagram, *one* arrow to show the direction of movement of carbon dioxide, between the alveolus and capillary.

arrow from plasma in capillary to air space in alveolus

1 mark

iii) Give the function of the red blood cell in this process.

carries oxygen away from alveolus

1 mark

iv) Describe *one* way that smoking tobacco might stop gaseous exchange from taking place properly.

Smoke weakens alveoli walls

Alveoli break down reducing surface area for gas exchange

Smoke contains carcinogens

Cancer reduces alveoli available for gaseous exchange

2 marks

c) Oxygen is important because it is needed for respiration.

i) Why is respiration important for all living things?

oxygen is used to release energy from food

1 mark

ii) Use the words from the box to write the word equation for respiration.

> **carbon dioxide   glucose   water**

glucose + oxygen → *water* + carbon dioxide

1 mark

SEG

3 The table shows the blood sugar levels of Rhys and Mair over a two hour period.

		Time									
		1.00 pm	1.15	1.30	1.45	2.00	2.15	2.30	2.45	3.00	3.15
Blood sugar levels (mg/ 100 cm³ blood)	Rhys	90	90	90	110	140	112	96	90	90	90
	Mair	90	90	90	128	170	192	200	208	200	180

↑
Meal eaten

a) At 1.30 p.m. both ate identical high carbohydrate meals. Briefly describe the effect the meal had on the blood sugar levels of Rhys and Mair:

Rhys: *the meal increased the blood sugar level; it started to decrease after 30 minutes*

2 marks

Mair: *the meal increased the blood sugar level; it started to decrease after 60 minutes*

2 marks

b) Name the medical condition that Mair is suffering from

diabetes

1 mark

c) Blood sugar level is controlled by two hormones, insulin and glucagon. The diagram below shows how this control is brought about.

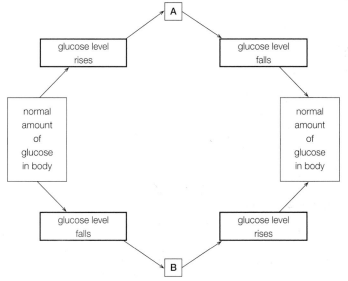

i) Name the organ that produces these two hormones.

pancreas

1 mark

ii) Explain fully what happens at:

A point **A** to lower the blood sugar level;

insulin is released by the pancreas

it causes liver cells to take in glucose and convert it to glycogen

2 marks

B point **B** to raise the blood sugar level.

glucagon is released by the pancreas; it causes liver cells to release glucose into the blood from stored glycogen.

2 marks

WJEC

4 The diagram below shows part of the human urinary system.

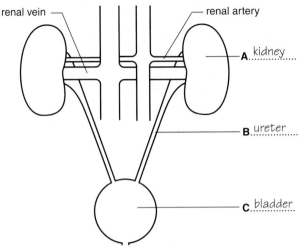

renal vein — — renal artery

A. kidney

B. ureter

C. bladder

a) Label parts **A**, **B** and **C** on the lines provided.

3 marks

b) Blood leaving a kidney contains less urea than blood entering it. The kidney has taken urea out of the blood. Describe what happens to this urea.

Urea is filtered out of the blood in the Bowman's capsule

It passes down the collecting duct into the ureter and from there into the bladder

From the bladder it is expelled in urine

3 marks

c) Give *two* ways in which a person can be treated if the kidneys stop working.

1   *dialysis*

2   *kidney transplant*

2 marks
Edexcel

5 a) The diagram shows the structure of the human eye.

i)   Add labels to the diagram. Choose words from this list.

ciliary muscles   cornea   iris   optic nerve   retina

3 marks

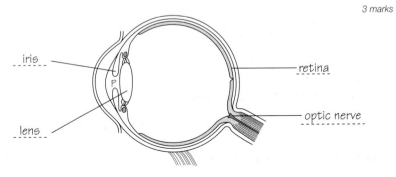

iris

lens

P

retina

optic nerve

ii) Write a **P** on the diagram to show where the pupil is.

1 mark

iii) Draw the lens in its correct position on the diagram.

1 mark

lens drawn as an oval shape behind the iris, connected to the ciliary bodies (round protrusions behind the iris)

b) i) A person moves from a dim room into bright sunshine. Describe the change in the appearance of her pupils.

pupils decrease in diameter

1 mark

ii) Some eye-drops prevent this change happening. Why is this dangerous to the eye?

too much light can damage the retina

1 mark

The eye is a sense organ that is stimulated by light.

c) Finish the chart by writing in the correct sense organ.
The first one has been done for you.

Stimulus	Sense organ
Light	*eye*
Chemical taste	*tongue*
Chemical smell	*nose*
Pressure	*skin*
Sound waves	*ear*

4 marks

d) Some stimuli can cause reflexes to take place. Which reflexes occur in the following examples?

i) When dust gets in your eyes.

blink

1 mark

ii) When pepper enters your nose.

sneeze

1 mark

iii) When food enters your windpipe.

cough

1 mark

e) The diagram shows a reflex arc.

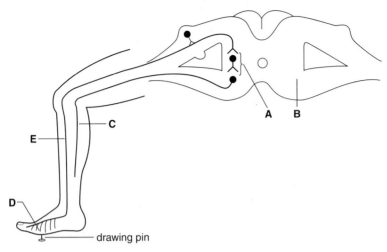

drawing pin

Complete the table by matching the letters of the diagram to the part of the reflex arc. The first one has been done for you.

Part of reflex arc	Letter
Relay neurone (nerve cell)	A
Receptor	D
Sensory neurone	E
Spinal cord	B
Motor neurone	C

3 marks

f) How do reflex actions help to protect our bodies from damage?

Because they do not require conscious thought, they can quickly deal with any harmful stimulus

1 mark
OCR

6 The diagram below shows a plan of the circulatory system. The blood vessels are labelled with letters.

Use the letters on the diagram to complete the sentences in the table below. The first one has been done for you.

Sentence	Letter
The blood vessel named the vena cava is	L
The blood vessel named the pulmonary artery is	M
The blood vessel carrying blood with the *most* oxygen is	B
The blood vessel carrying blood with the *most* glucose after a meal is	J
The blood vessel carrying blood with the *least* urea is	I
The blood vessel containing blood at the *highest* pressure is	C

5 marks

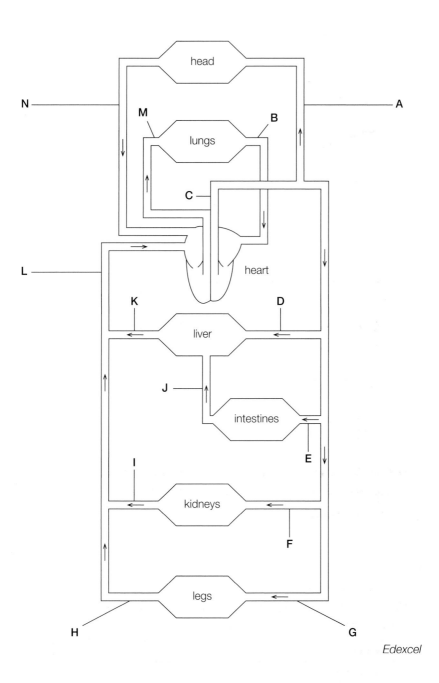

N

A

M

B

head

lungs

C

heart

L

K

D

liver

J

intestines

E

I

kidneys

F

H

legs

G

Edexcel

PART THREE

KS3

1 The following diagram shows the cross-section of a flower.

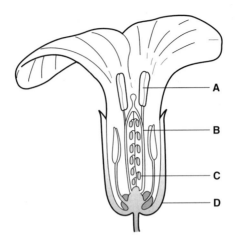

a) Parts of the flower are labelled with the letters **A**, **B**, **C**, **D**. Complete the table by writing the correct letter in the correct box.

Part of flower	Letter
Ovary	*B*
Ovule	*C*
Sepal	*D*
Stamen	*A*

4 marks

b) What is the function of the part labelled '**A**'?

to produce pollen

1 mark

2 The diagram below is a plant cell.

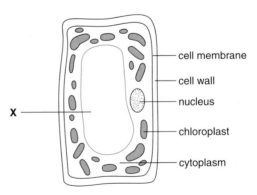

a) Which two named parts occur in plant cells, but not in animal cells?

1   *cell wall*

2   *chloroplast*

2 marks

b) In which *named* part is the genetic information contained?

nucleus

c) Which *named* part is involved in absorbing light energy for photosynthesis?

chloroplast

1 mark

d) Write down the name of part **X**.

vacuole

1 mark

e) In which part of the plant would you find a cell like the one shown in the diagram?

in the centre of a root ☐   in the lower surfaces of a leaf ☐

near the upper surface of a leaf ☑   near the surface of a root ☐

1 mark

3 Carbon dioxide is used by plants in photosynthesis. The amount of carbon dioxide that a tree takes in each day varies over a whole year. This is shown in graph A.

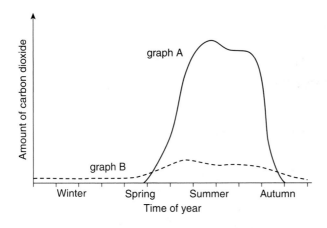

a) i) Write down two reasons why photosynthesis happens more quickly in the summer.

it is warm; there is more light

the days are long; the tree has more leaves

2 marks

ii) Which two substances are produced in photosynthesis?

glucose (starch) and oxygen

2 marks

b) Deciduous trees lose all their leaves in the autumn. Graph A shows the carbon dioxide intake of a deciduous treee. How can you tell from Graph A that this tree is deciduous?

photosynthesis stops in the autumn and starts in the spring

1 mark

c) The amount of carbon dioxide produced by the tree each day also varies over a whole year. This is shown in Graph B. Write down the name of the process which produces carbon dioxide.

respiration

1 mark

KS4

1 a) Give a definition of osmosis.

The diffusion of water molecules from a region of high water concentration to a region of lower water concentration across a semi- or selectively permeable membrane OR The diffusion of water molecules from a region of low substrate concentration to a region of higher substrate concentration across a semi- or selectively permeable membrane

4 marks

b) What is the difference between osmosis and diffusion?

osmosis only refers to the diffusion of water

1 mark

c) The diagram shows a section through a root hair cell.

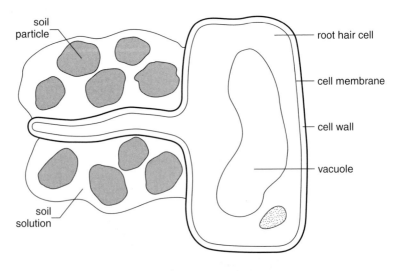

Explain how water passes from the soil solution to the root hair cell.

Because there is a relatively high water concentration outside the root hair, water diffuses into the root by osmosis down its concentration gradient. The water concentration is relatively low inside the root hair cell because there is a high concentration of minerals and sugars dissolved in the cytoplasm.

2 marks

d) The table shows the concentration of certain mineral ions in the soil solution and in the root hair cells:

	Concentration mmol dm^{-3}		
	Potassium	**Sodium**	**Chloride**
Soil solution	0.1	1.1	1.3
Vacuole of root hair cell	93.0	51.0	58.0

i) What is the evidence, from this table, that uptake of mineral ions from the soil is by active support?

The root hair cell must have expended energy to pump minerals back into the cell against their concentration gradient.

1 mark

ii) Name a part of the human body where active transport takes place.

kidney

1 mark

WJEC

2 A green plant was placed in a dark cupboard. After 24 hours, some of the leaves were tested for starch. No starch was found in any leaf. The same plant was then placed in sunlight. One leaf, **Q**, was treated as shown in the diagram below. After a further 24 hours, leaves **P** and **Q** on the plant were tested for starch.

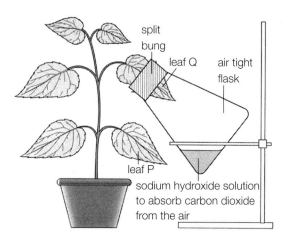

split bung

leaf Q

air tight flask

leaf P

sodium hydroxide solution to absorb carbon dioxide from the air

a) Why was there no starch in the leaves after the plant had been kept in the dark cupboard for 24 hours?

The plant needs light to produce starch by photosynthesis. Without any light, no new starch could be produced. Any starch there previously would have been transported away or converted to glucose for use in respiration.

2 marks

b) i) Name the substance in green leaves which helps plants to make glucose.

chlorophyll

1 mark

ii) Name the gas produced by green leaves when they make glucose for starch production.

oxygen

1 mark

c) The diagrams below show the results of starch tests on discs taken from leaves **P** and **Q**. The diagrams below show the parts of the leaves from which the discs were taken.

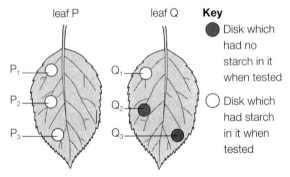

Explain the reasons for the differences between discs **P3** and **Q3** and between discs **P2** and **Q2**.

Disc Q3 had no access to carbon dioxide. Disc P3 had access to carbon dioxide. Plants need carbon dioxide to make starch.

Disc Q2 had no access to light (it was under the bung). Disc P2 had access to light. Plants need light to make starch.

3 marks
Edexcel

3 The diagrams show the details of an experiment with an auxin.

a) Why was petiole **B** covered in petroleum jelly that did not contain an auxin?

as a control

1 mark

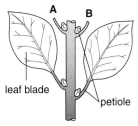 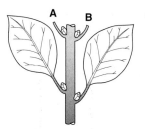 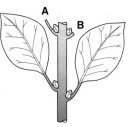

Stage 1:	Stage 2:	Result:
Leaf blades removed at **A** and **B**	Petiole **A** coated with an auxin in petroleum jelly. Petiole **B** coated with petroleum jelly only	After a few days petiole **B** has fallen off

b) Using the information from the experiment, explain how leaf-fall might be controlled by the plant.

Auxin prevents leaf fall

Petiole B fell off in response to no auxin

2 marks

c) When a gardener removes the shoot tip of a shrub, side branches start to grow out from the main stem. Explain why this happens.

The auxin produced by the shoot tip inhibits growth of side buds. When the shoot tip is removed, side buds can grow.

2 marks

SEG

PART FOUR

KS3

1 Jane keeps chickens. During the night she locks them in a shed. During the day, they are free to run around her garden. The table below contains information about the chickens.

Name of chicken	Sex	Number of eggs laid last year
Molly	female	160
Emily	female	190
Skinney	female	150
Lucy	female	200
Elvis	male	–

a) Jane wants her hens to lay as many eggs as possible. By changing their food and water, she can give them the best recommended balanced diet. Suggest *two* other ways she could alter the chickens' living conditions to increase the number of eggs laid.

239

keep them free of disease

keep the shed at the correct temperature

give them more light during the winter

keep them indoors

keep them or the shed clean

2 marks

b) Jane decides to use selective breeding to get chickens which lay a large numbers of eggs each year.

i) Which hen should Jane use as a mate for Elvis at the start of her breeding programme?

Lucy

1 mark

ii) How should Jane continue to choose hens from the offspring to continue her breeding programme?

she should choose those which lay most eggs

1 mark

c) Jane's chickens are all the same breed: Rhode Island Red. Their living conditions do *not* effect the colour of their feathers. What *does* determine the colour of their feathers?

feather colour is inherited

feather colour is controlled by genes, chromosomes or DNA

1 mark

d) Eggs contain all the food and water required by a developing embryo until it hatches. The eggshell has lots of tiny holes in, making it porous. Why is the eggshell porous?

to let oxygen in

to let carbon dioxide out

1 mark

KS4

1 Wheat is an important food crop and new varieties have been produced by *selective breeding* to improve yield and *reduce waste*.

The following table shows some of the features of five wheat varieties produced during this century.

a) What is meant by the term *'selective breeding'*?

Identify the required characteristics in individual organisms. Breed these organisms together repeatedly

2 marks

Variety	Year of first use	Mean height of plant (cm)	Grain yield (tonne per hectare)
Little Joss	1908	142	6.0
Holdfast	1935	125	6.0
Capelle Desprez	1953	110	6.7
Maris Huntsman	1972	106	7.5
Norman	1980	84	8.7

b) Describe *two* clear trends shown in the table.

1   *plants become shorter*

2   *plants become more productive*

2 marks

c) Explain how the wheat breeders can claim to have *'reduced waste'* in wheat crops.

plants have become shorter, but there is more grain, therefore less of the plant is thrown away

2 marks

Edexcel

2 a) The genetic diagram shows how height is inherited in peas.

T represents the allele for tallness.

t represents the allele for shortness.

i)   Using words from the list, fill in the spaces in the genetic diagram. Each word may be used once, more than once or not at all.

**dominant   gametes   genotype   heterozygous   homozygous
phenotype   recessive**

Parental *phenotype*	tall	short
Parental *genotype*	*TT*	*tt*
gametes	all (T)	all (t)
F₁ *genotype*	all *Tt*	
F₁ *phenotype*	all tall	

F₁ is written as F_1 *genotype* — all *Tt*; F_1 *phenotype* — all tall

3 marks

ii)   Complete this paragraph using words from the list above.

When pea plants have two identical alleles for tallness or shortness, i.e. *TT* or *tt*, they are

homozygous

When they have two different alleles, i.e. *Tt*, they are

heterozygous

2 marks

iii)   Which allele is recessive? Explain your answer using the results of this cross.

241

*The allele for short, t, is recessive. This is because its effect is masked by T
(the allele for tallness) when they are together within an individual.*

2 marks

b) Haemophilia is a sex linked characteristic.

People with haemophilia often bleed because their blood does not clot easily.

It is caused by a recessive allele (*h*) on the X chromosome. (X^h)

People with the dominant allele (*H*) have normal clotting of blood. (X^H)

The diagram shows how the allele for haemophilia is inherited.

i) Complete the checkerboard to show the genotypes of the children.

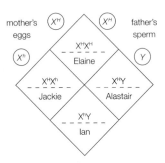

4 marks

ii) Explain why the mother is described as a carrier of this condition.

*Because she has the allele for haemophilia, but does not develop the
disease herself. This is because she is heterozygous and the normal allele
masks the effect of the haemophilia allele*

2 marks

iii) Which children have haemophilia?

Ian

1 mark

iv) Which children would be carriers?

Jackie

1 mark

OCR

3 In humans, the body cells contain 23 pairs of chromosomes. The sex
chromosomes form one of these pairs. The diagram below shows the sex
chromosomes of female and male individuals.

Sex chromosomes from two individuals

a) The sex chromosomes are usually represented by the symbols **X** and **Y**.

i) Using these symbols, state which sex chromosomes are present in:

A a female; *XX*

B a male. *XY*

2 marks

ii) State which sex chromosomes are present in:

A an ovum; *X only*

B a sperm. *X or Y*

2 marks

b) Use the symbols and a genetic diagram to show how offspring of either sex are formed.

Parents	Male	Female
	XY	*XX*
Gametes	*X or Y*	*X or X*
Offspring		

	X	*Y*
X	*XX*	*XY*
X	*XX*	*XY*

Offspring 50% male, 50% female

2 marks

SEG

4 In mice, the allele for black fur is dominant over the allele for brown fur. The allele for black fur can be represented by '*B*' whilst '*b*' represents the allele for brown fur.

a) i) What is the genotype of a brown mouse?

bb

1 mark

ii) What are the possible genotypes of a black mouse?

BB or Bb

2 marks

b) A pure breeding (homozygous) black female mouse mated with a pure breeding brown male.

i) What is the genotype of all the offspring?

Bb

1 mark

ii) What is the phenotype of all the offspring?

Black

1 mark

c) A male and female mouse, both with the genotype *Bb*, mate. The female produces a litter of 12. Use a genetic diagram to show the genotypes of the offspring and to predict the numbers of brown and black offspring.

Parent phenotype   *Black*   *Black*

Parent genotype   *Bb*   *Bb*

Fertilisation and offspring genotype

	B	*b*
B	*BB*	*Bb*
b	*Bb*	*bb*

Offspring phenotype 75% Black, 25% Brown. Therefore, from a litter of 12, you would expect nine black mice and three brown mice

4 marks
SEG

5 The map shows:

■ Densely populated industrial areas

○ All normal pale forms

● All mutant dark forms

◑ Combinations of both forms

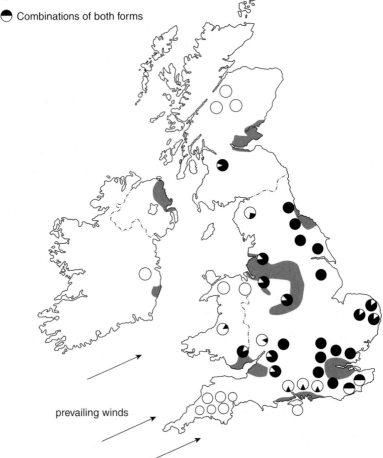

prevailing winds

the most densely populated industrial areas;

the frequency of pale and dark forms of the peppered moth;

the direction of the prevailing winds in the British Isles.

Peppered moths usually rest on trees covered with lichen, and they are preyed upon by many birds. In areas of low air pollution the lichen on trees is usually pale in colour. In areas of high air pollution the lichen turns black.

a) i) State a pattern of the distribution of the mutant dark form shown on the map.

The mutant form is concentrated around big cities

1 mark

ii) Suggest a reason for your pattern.

pollution turns the trees dark, therefore any pale forms would be seen by predators and eaten.

1 mark

b) The dark form of peppered moth developed after a *mutation* in the pale form. What is a *mutation*?

a change in a gene

1 mark

c) Using the idea of natural selection explain why the dark form of the moth is restricted to the areas shown.

The dark form of the moth is camouflaged in polluted areas. The pale form is not camouflaged as effectively. The dark form will therefore survive to reproduce. The 'dark' genes will be perpetuated into the next generation. The population will eventually be made up completely of dark forms. Any dark form moving into an unpolluted area will be predated.

4 marks
SEG

PART FIVE

KS3

1 Primroses and violets grow in a wood. They are small plants. One night there was a storm, and some of the trees fell down. They were taken away the next day.

a) During the next year, the number of primroses and violets growing in the wood had increased. Give two reasons why.

more light available for growth

more room available to grow

more seeds germinated

2 marks

b) After the trees had been removed, the number of birds living in the wood decreased. Give *two* reasons why.

fewer places to nest

fewer places to perch or rest

less food

fewer hiding places

2 marks

2 The diagram below shows part of a woodland food web.

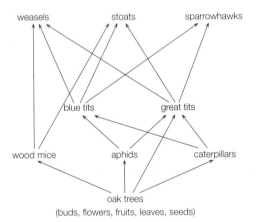

oak trees
(buds, flowers, fruits, leaves, seeds)

Use only the information given in the diagram, answer questions a) and b).

a) Write down the name of one herbivore and one omnivore.

herbivore *wood mouse, aphid, caterpillar*

omnivore *great tit*

2 marks

b) The population size of the blue tits decreases. This affects the number of great tits.

i) Suggest one reason why the population size of the great tits may increase.

less competition for caterpillars or aphids

1 mark

ii) Suggest one reason why the population size of the great tits may decrease.

weasels or stoats or sparrowhawks eat more of them

1 mark

iii) If the number of blue tits decreased, why would this affect the sparrowhawks more than the stoats?

Because sparrowhawks only eat blue tits and great tits, whereas the stoats

also eat wood mice

1 mark

c) The arrows in a food web show the direction in which energy is flowing through an ecosystem. However, when a weasel eats a wood mouse, most of the chemical energy stored in the wood mouse does not get stored as chemical energy in the weasel. Explain why.

Lost as thermal energy or by respiration

Lost in undigested material or waste

Lost through excretion

2 marks

d) The diagram below shows four different pyramids of numbers.

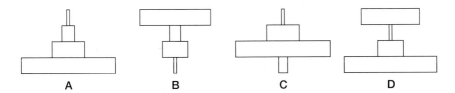

| A | B | C | D |

Which of the diagrams is most likley to be the pyramid of numbers for the following food chain:

oak trees → aphids → blue tits → sparrowhawks

C

1 mark

KS4

1 The diagram below shows a woodland food web.

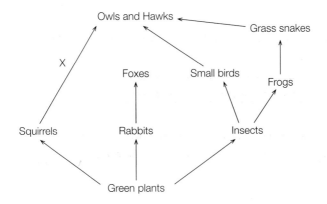

a) Using the information above:

i) Name a herbivore.

squirrel, rabbit, insects

1 mark

ii) Name a carnivore.

fox, owl, hawk, frog, grass snake, small birds

1 mark

iii) What is the source of energy for *all* the living organisms in the food web?

the Sun

1 mark

iv) Name *two* effects on the food web if all the frogs were killed.

the grass snakes would die or have to eat different food

less insects eaten and population would increase

more food for small birds so population increase

1 mark

v) Explain fully what the arrow at **X** means.

squirrels eaten by owls and hawks

b) A

| Birds |
| Insects |
| Tree |

B

| Birds |
| Insects |
| Tree |

Which of the diagrams given above represents:

i) a pyramid of numbers *A*

ii) a pyramid of biomass? *B*

2 marks
WJEC

2 a) The diagram shows what happens to each 1000 kJ of light energy absorbed by plants growing in a meadow.

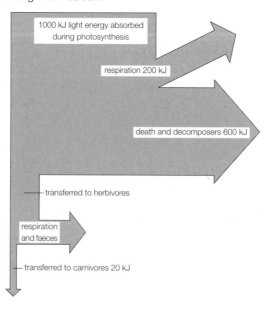

1000 kJ light energy absorbed during photosynthesis

respiration 200 kJ

death and decomposers 600 kJ

transferred to herbivores

respiration and faeces

transferred to carnivores 20 kJ

Use information from the diagram to calculate:

i) how much energy was transferred to herbivores:

1000 − 200 − 600

200 kJ

1 mark

ii) the percentage of the energy absorbed during photosynthesis that was eventually transferred to carnivores. Show your working.

(200 ÷ 1000) × 100

= 2%

1 mark

b) The table gives the energy output from some agricultural food chains.

Food chain	Energy available to humans from food chain (kJ per hectare of crop)
cereal crop → humans	800 000
cereal crop → pigs → humans	90 000
cereal crop → cattle → humans	30 000

Explain why the food chain *cereal crop → humans* gives far more energy to humans than the other two food chains.

Energy is lost from a food chain at each level. For example, through respiration and waste.

By eating the cereal crop directly, humans gain that energy which would have been lost by the pigs or cattle.

3 marks

c) The amount of energy available to humans from the food chain *cereal crop → pigs → humans* can be increased by changing the conditions in which the pigs are kept.

Give *two* changes in conditions which would increase the amount of energy available. In each case explain why changing the condition would increase the available energy.

Change of condition 1 *keep the pigs warm*

Explanation *pigs waste less energy keeping warm*

Change of condition 2 *limit pigs' movement and provide with food*

Explanation *pigs waste less energy moving and looking for food.*

4 marks

NEAB

2 The diagram shows part of the nitrogen cycle.

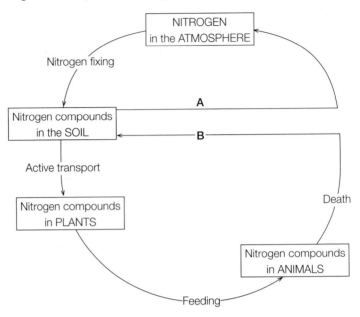

a) What processes are represented by the arrows **A** and **B**?

A   *denitrification*

B   *nitrification*

2 marks

b) Explain *three* steps involved in how nitrogen compounds in plants become nitrogen compounds in animals.

1   *animal eats the plant*

2   *the plant is digested and the products absorbed*

3   *nitrogen compounds transported to the cells*

3 marks

NICCEA

3 The table shows birth rates and death rates for some countries.

Country	Annual birth rate per 1000 population	Annual death rate per 1000 population
China	20	8
USA	15	9
India	34	15
Britain	12	12
Sweden	12	11

a) i)   Which country shows the greatest annual rate of increase in its population?

India

1 mark

ii) Which country shows *no* annual increase in its population?

Britain

1 mark

iii) Which country shows the smallest annual rate of increase in its population?

Sweden

1 mark

b) Give *two* major problems faced by a country with a large increase in its population.

lack of food

increased chance of disease

insufficient infrastructure

unemployment/economic disadvantage

2 marks
SEG

Index

NOTES

NOTES

NOTES

NOTES

NOTES

NOTES

NOTES